T0177502

Male Choice, Female Competition, and Female Ornaments in Sexual Selection

Male Choice, Female Competition, and Female Ornaments in Sexual Selection

Ingo Schlupp

Department of Biology, University of Oklahoma, USA

OXFORD

UNIVERSITY PRESS

OXFORD

UNIVERSITY PRESS

Great Clarendon Street, Oxford, OX2 6DP,
United Kingdom

Oxford University Press is a department of the University of Oxford.
It furthers the University's objective of excellence in research, scholarship,
and education by publishing worldwide. Oxford is a registered trade mark of
Oxford University Press in the UK and in certain other countries

First Edition published in 2021

Impression: 2

Published in the United States of America by Oxford University Press
198 Madison Avenue, New York, NY 10016, United States of America

British Library Cataloguing in Publication Data

Data available

Library of Congress Control Number: 2020951911

ISBN 978–0–19–881894–6

DOI: 10.1093/oso/9780198818946.001.0001

Printed and bound by
CPI Group (UK) Ltd, Croydon, CR0 4YY

Contents

Preface

Writing this book was a deeply personal endeavor for me. Books have their own personal history and an agenda, mainly because authors have them. Like many other projects, I started this book on a dark and stormy night in the field many years ago. Without the opportunity generously provided by my university to work on this during a sabbatical, I would have never finished. I am very grateful for that. I am also grateful to several funding agencies that made possible the work that I refer to throughout the book: Caribaea Initiative, German Science Foundation (DFG), German Academic Exchange Service (DAAD), Alexander v. Humboldt Foundation, National Geographic, and the National Science Foundation (NSF).

In addition to being personal, books are also written at a certain time and inevitably must reflect this. As I was working on this book the climate change crisis got worse by the day, with very little good news. Why even work on a book on mate choice when the planet is going down the drain? And how can I sit in my office at home and work on this, with the world in Covid-19 turmoil around me? I have no good answer, but I am simply not ready to give up hope.

This book, like all books, I believe, reflects my limitations and biases. I feel, for example, more comfortable writing about fishes, because that is the taxonomic group I mostly work with scientifically. That is part of the reason for my reliance on fish examples. But also, fishes are the most speciose group of vertebrates and more variable in almost all traits than other vertebrates. I admit, though, that I am probably less well informed about other taxa, although I tried very hard to read and cite the literature widely.

In addition to these intrinsic biases, there are external constraints to what an author is doing. There are deadlines, word limits, and reviews. All of this leads to an imperfect product and an emotional rollercoaster for the author. I am sure that I will notice errors and flaws in my book the day after I submit it and lose control over it, but, eventually, an author has to let go and live with the consequences. Just like most authors, I have a love–hate relationship with my book. It was great fun to write it and to intellectually engage with a topic, but it is also frustrating as I struggle for words (English is not my first language, although that is not meant to be an excuse). Sometimes I fear I am producing an unreadable word salad that no one will care to read. Of course, I worry about whether my ideas will be well received or not. Or if people will laugh at the errors I made or the emphasis I choose. Putting these anxieties aside is not easy. Yet, my biggest dream and hope for the book is that it will be outdated and obsolete in a few years.

Writing a book was primarily a process and the ideas I had early on changed substantially over time. This may be the best thing about the process of writing this text. It was deeply gratifying to be able to talk to people about my ideas and thoughts. Most colleagues were very gracious with their time and engaged in discussions about questions I had for them. Most academics are very friendly, kind, and have preserved the curiosity and open-mindedness that got them into research. I am very grateful for that.

Maybe the worst thing about a book format is that it pretends to be linear, one page following the other. But knowledge is not linear. Everything is connected, sometimes in unexpected, even unpredictable, but always exciting ways.

A book like the present one is always a long time in the making. I had support from countless people, for which I am immensely grateful. I dedicate this book to my family, in particular my wonderful wife,

Andrea. Her kind help and patience at every step of the process made this project possible. Our four wonderful children supported me too, each in their own unique way. Without them my life and my work would have a lot less meaning. I am also grateful to my parents and to Andrea's parents. Both Andrea and I lost our mothers while I was working on this book, so this is for them too.

I was able to run ideas (including all the stupid ones), chapters, and thoughts by a lot of people. They all helped me to think a little more clearly and I am very grateful for their time and effort. I feel honored and lucky to be part of such a great community. But of course, all errors and shortcomings of this work are solely mine.

Maybe most importantly, I have to acknowledge the many great mentors I had. They helped me through the many crises that accompany a career in academia and gave me hope. Many come to mind, but four stand out: Jakob Parzefall, Manfred Schartl, Uli Reyer, and Michael J. Ryan. I hope I can pay a little of this forward. Finally, I am grateful to all my students, present and past. They all made incredibly important contributions to my thinking and I often think that I am learning more from them than they are from me.

Oxford University Press has a fine group of editors and I am indebted to Ian Sherman, Bethany Kershaw, and Charles Bath for accompanying me patiently as I missed deadline after deadline along the long road to the final book. Special thanks go to Andrea Schlupp, who bravely read all chapters of the book and provided comments. Several others read at least one chapter and provided feedback: Claire Doutrelant, Amber Makowicz, Rudy Riesch, and Michael J. Ryan.

Some colleagues very kindly and generously gave permission to publish their artwork: Theo Bakker, Anders Berglund, Deike Lüdtke, David Funk, Sarah Lipshutz, Manfred Milinski, Florian Schiestl, Andreas Svensson, Katherine Swiney, Kazunori Yoshizawa, and Rudy Riesch.

More thanks are due to the many people listed here for fruitful discussion or help of some kind:

Ingrid Ahnesjö
Theo Bakker
Charles Bath

Waldir Miron Berbel-
    Filho
Anders Berglund

Ari Berkowitz
Niko Besnier
Alexis Billings
Warren Booth
Stefan Bräger
Sarah Bush
David Buss
Reagan Cannetti
Christa Chancellor
Rickey Cothran
Darren Croft
Claire Curry
Innes Cuthill
Etienne Danchin
Karen deJong
Tobias Deschner
Nick DiRienzo
Marcel Dorken
Claire Doutrelant
Matt Dugas
Melissa Emery
    Thompson
Courtney Fitzpatrick
Doug Gaffin
Antje Girndt
Cari Goetz
John Goodwin
Johana Goyes Vallejos
Rosemary Peter Grant
Palestina Guevara-Fiore
Piers Hale
Daniel Hanley
Rachel Hazlitt
Rob Heinsohn
Jenn Hellmann
Christie Henry
Marielle Hoefnagels
Kerstin Johannesson
Kristine Johnson
Peter Kappeler
Bethany Kershaw
Ellen Ketterson
Abby Kimmitt
Willow Lindsay
Sara Lipshutz
Deike Lüdtke
Martine Maan
Constantino Macías
    Garcia

J. P. Masly
Carolyn Mead Harvey
Tamra Mendelson
Marcela Méndez-Janovitz
Manfred Milinski
Tania Munz
Hazel Nichols
Sabine Nöbel
Joyce Parga
Heike Pröhl
Eveleen Richards
Mike Ritchie
Rodet Rodriguez
Gil Rosenthal
Christian Rutz
Michael J. Ryan
Scott Sakaluk
Jutta Schneider
Gordon Schuett
Kristina Sefc
Maria Servedio
seven anonymous
    reviewers for Oxford
    University Press
Ian Sherman
Atsushi Sogabe
Montrai Spikes
Staats- und
    Universitäts-
    Bibliothek Hamburg
    Carl von Ossietzky
Kyle Summers
Lindsey Swierk
Michael Taborsky
Katie Taylor
the great people at the
    OU Bizzell Library
Timo Thünken
Ralph Tiedemann
Claus Wedekind
Gary Wellborn
Klaudia Witte
Kazunori Yoshizawa

Ingo Schlupp
Professor of Biology
Department of Biology
University of Oklahoma,
USA

Glossary

The glossary is intended to provide a quick definition of important terms used throughout the book. They are only meant as an entry point.

Adult sex ratio (ASR) ratio of number of adult males to females in a population.

Anisogamy differences in size between gametes. Large gametes are typically eggs, small gametes are typically sperm.

Bateman's principle based on anisogamy males have more mating partners than females and higher variance in reproductive success.

Binary choice test method to test for mating preferences with one choosing individual that is given a choice between two stimuli.

Kuhnian revolution a major shift in paradigms in a scientific discipline.

Lek small area where courters display to choosers without providing any resources.

Major histocompatibility complex (MHC) a group of genes that are important in immune defense. More diversity is thought to be beneficial.

Operational sex ratio (OSR) ratio of number of males to females in a population ready to mate.

Pleiotropy gene that affects multiple phenotypic traits.

Potential reproductive rate (PRR) theoretical offspring production in a period of time if unconstrained by availability of mating partners.

Sex roles based on Bateman's principle males are thought to benefit from mating with multiple partners, while females benefit from selecting a few partners. There are four character states for this: a male with male-like behavior, a female with female-like behavior, a male with female-like behavior, and a female with male-like behavior.

Sexual conflict this can arise when the two mating partners have different reproductive optima.

Sexual dimorphism differences between two sexes.

Sperm depletion male sperm availability can go down as a function of multiple copulations.

CHAPTER 1

Sexual Selection, Mate Choice, and Competition for Mates

1.1 Brief outline of the chapter

Well over a century ago Charles Darwin redefined biology and introduced the theory of natural selection. One of the problems he encountered was the existence of traits, mainly in males, that seemed to defy the principles of natural selection: they did not aid its bearers in survival and were often outright detrimental. Darwin solved this conundrum by introducing sexual selection. Other than in natural selection where all individuals compete with each other for survival and reproduction, in sexual selection individuals within each sex compete with each other for reproduction. In the original formulation of the principle, Darwin recognized two mechanisms for this: males would compete with each other for access to females, and females would choose mating partners of their preference.

In this opening chapter I want to introduce the topics to be covered in the book, define some basic terms that we will need to understand the subject matter, and define the questions to be asked. My aim for this book is to summarize our growing, yet still comparatively limited empirical knowledge and theory, and to provide suggestions for future research. What interest me most are the relationships between the four forms of sexual selection and their consequences.

1.2 Sexual selection

Sexual selection is one of the most important forces in evolution. Charles Darwin introduced the theory in his revolutionary work *On the Origin of Species* in 1859 (Darwin, 1859). But as Darwin was working on his all-encompassing theory of natural selection, he came across many examples of traits that were difficult to explain within his new framework. They were mainly found in males and clearly were detrimental to their bearers. How could selection favor traits that did not contribute to survival? The solution he proposed in 1859 and then in much more detail in another book, *The Descent of Man, and Selection in Relation to Sex* in 1871 (Darwin, 1871) was sexual selection. He suggested that males may leave behind more offspring if they had more mating opportunities than others, either because they are favored by females or because they succeeded in competition with other males. We all know impressive examples of these two key mechanisms of sexual selection, female choice and male competition. In an example of the latter, gigantic males fight over access to females in elephant seal (*Mirounga angustirostris*) colonies. Males compete with each other over harems of females (Leboeuf, 1972). Owing to this, males have evolved a striking sexual dimorphism and are several times the size of the females. Intuitively, most biologists will likely attribute the existence of ornaments and also sexual dimorphism to sexual selection. Later in Chapter 7 I will discuss a few examples where this is misleading. Also, the sexual dimorphism in humans, which is usually attributed to sexual selection, may also be explained by hormonal and developmental effects (Dunsworth, 2020). This idea is by no means widely accepted but provides an example that we need to be cautious when we generalize. Male competition plays out within one sex (intrasexual). On the other hand, in female choice, females in numerous species pick their favorite males out of a line-up of

Male Choice, Female Competition, and Female Ornaments in Sexual Selection. Ingo Schlupp, Oxford University Press (2021). © Ingo Schlupp (2021).
DOI: 10.1093/oso/9780198818946.003.0001

males who are displaying byzantine dances while showing off ornamental traits, such as colorful plumage, song, smell, and motion patterns among others. Even the giant sperm found in some *Drosophila* are considered an ornament (Lüpold et al., 2016). A classic example is the tail of the peacock (*Pavo cristatus*). This particular example goes back all the way to Darwin, who agonized over the peacock's tail because it was impossible to explain its evolution with natural selection alone. He complained to Asa Gray in 1860 that the "sight of a feather in a peacock's tail...makes me sick" (Hiraiwa-Hasegawa, 2000). While natural selection is somewhat blind to sex (Sayadi et al., 2019), female choice is intersexual and plays out between the sexes. This is all happening to allow individuals to reproduce. The peacock's tail is also relevant as an example of indirect benefits (Petrie, 1994). Many courtship displays are multimodal (Rosenthal, 2017), and involve multiple sensory channels (Candolin, 2003; Partan and Marler, 2005). A great example of this is the courtship display of the Gunnison sage grouse (*Centrocercus minimus*), which blends color, motion patterns, and sound (Gibson et al., 1991).

Clearly, finding a partner for reproduction is likely the most important decision any organism ever makes. This is how genes are passed on to the next generation. This applies to direct offspring, and to some degree to offspring of close relatives via inclusive fitness. And, unsurprisingly, mate choice is very complex and complicated (Rosenthal, 2017). It is also a highly interactive process with many different phases, from recognizing a suitable mate to the actual fertilization of gametes. Furthermore, it often seems like an agreeable, almost cordial process, but in actuality is usually fraught with conflict because only rarely do the partners have identical fitness returns (Chapman et al., 2003).

The original definition of sexual selection provided by Darwin has been subject to intense debate and several other definitions have been suggested (Alonzo and Servedio, 2019). All the definitions agree that the hallmark of sexual selection is that it has effects within a sex and highlight differential reproductive success as the key mechanism (Alonzo and Servedio, 2019). A problem with many definitions is that they rely on binary sex roles and assume that we can always assign a male and female role. Sexual

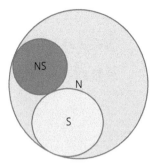

Figure 1.1 Selection in the wild. N equals natural selection, S equals sexual selection, NS equals combined natural and sexual selection (redrawn from Andersson, 1994).

selection can be viewed as a subset of natural selection. Natural selection works on both sexes, although it can work on them separately (Sayadi et al., 2019). To visualize the relationship of natural selection and sexual selection, perhaps one could think of natural selection as the outer layer of an onion. Beneath lies sexual selection, with its four components, female and male mate choice, and male and female competition. These elements and their interplay will determine the reproductive success of an individual (Figure 1.1). One might think that sexual selection theory has been studied enough and is becoming boring, but the opposite is true (Lindsay et al., 2019).

Within sexual selection, two components have received much theoretical and empirical attention: males competing over access to mating opportunities with females and females choosing among competing males. This view, however, appears to leave out some important elements, such as female competition and male mate choice. Male mate choice in particular has been "flying under the radar" for a long time, but recent years have seen a small surge of studies that address what I call the *orphans* of sexual selection theory: male choice and female competition (Figure 1.2). These two additional elements have been somewhat studied since Darwin's time, but are far less well documented and understood than female choice and male competition, the two major elements of sexual selection introduced by Darwin.

Sexual selection theory has an interesting history that may provide some more general lessons to be

Figure 1.2 Complete matrix of sexual selection.

learned. Male competition was relatively quickly accepted because this aspect apparently was a good fit for Victorian scientists and society (Milam, 2010; Richards, 2017). The notion of a mainly passive female and a knightly competition for them was easier to rationalize than active female choice. Actually, from the very beginning female choice was a tough sell, and Darwin experienced major pushback from his colleagues, including Wallace, who was his ally on natural selection (Richards, 2017). Even Darwin himself, perhaps being a child of the Victorian era, had problems with female choice: he conceived the idea of male competition around 1842, but female choice was not formalized until over a decade later around 1856 to 1858 (Richards, 2017). Darwin uniquely also assigned human men the ability to choose females. This was in contrast to other parts of his theory but has led to the peculiar situation that male mate choice has been a natural topic for evolutionary psychology for a long time—leading to a rich literature on humans, while it was somewhat ignored in other animals.

Female choice finally took a central role in research much later—actually the delay was over a century. To me the late blooming of female choice is an important cautionary tale, showing how societal context influences research and can act as blinders for research. In an early paper, Zuk (1993) summarizes a feminist perspective of sexual selection research and points out how the growing number of female scientists has changed research paradigms, interpretations, and language use to better reflect societal changes. The concept of sex

roles and the stifling effect this has had on the development of the field has also been critiqued (Ah-King, 2013a, 2013b; Ah-King et al., 2014). A proposed solution to this is to use the term "chooser" (Rosenthal, 2017) for any individual that is selective in mate choice. It is important to note that the emphasis is on individual behavior, not sex roles that apply to populations or species. A recent book on mate choice (Rosenthal, 2017) used this terminology and replaced male and female with neutral terms, such as "chooser" for individuals that select mates and "courter" for individuals that compete for mates and display to choosers in some form. This circumvents loaded terminology such as "sex-role reversal" for mating systems in pipefishes, where males are choosers (Berglund and Rosenqvist, 2003).

This struggle with terminology reflects a very general phenomenon. It is very important here to realize that language is very powerful, and in itself may influence how we understand important concepts such as time (Boroditsky, 2001). For example, mate choice is sometimes used as shorthand for female choice (Dargent et al., 2019), effectively removing male mate choice from consideration. Generally speaking, there is always a very tight connection between the science that is done and the current condition of society. This can take many different shapes and forms, such as directly outlawing certain research, like stem cell research using cell lines originating from abortions in the United States based on moral and religious grounds mainly outside of the scientific community, but with clear input from the scientists (McLaren, 2001; Nisbet, 2005). In the United States, legislation banning stem cell research was the outcome of an intense public debate over these values (Ho et al., 2008). Other countries, based on their values and laws, allow stem cell research. This shows how the values of societies, and the resulting governments and administrations, set the framework and the limits for what scientists are allowed to do. Some research, such as studies on gun violence in the United States, is not outright forbidden but may not be funded by public monies. This ban goes back to the so-called Dickey Amendment, which was introduced by US House Representative Jay Dickey (Jamieson, 2013), a Republican from Arkansas, in response to a 1993 paper (Kellermann et al., 1993) showing a connection between the presence of a gun in a home and the risk of homicide. Subtler and always well intended, but no less consequential, are the policies set by funding agencies that steer research in certain directions by allocating or taking away funds. But even without outside influences the science actually conducted is obviously influenced by the relevant societal context. This requires a constant engagement of scientists and society on many important issues. A recent example would be the use of cloning technology on humans (Liu et al., 2018), or the ongoing debate on gene-editing techniques using clustered regularly interspaced short palindromic repeats (CRISPR)-Cas9. These examples also highlight an important aspect of the feedback loop between society and the scientific community, namely that oftentimes scientific discovery happens a long time before a meaningful discussion of its implications can be held. While some of these discussions are held in public, and sometimes loudly, many discussions and disputes are more internal and never reach a general audience.

Does this also apply to aspects of sexual selection? One reason for the relative lack of studies on male mate choice and female competition could be that based on Darwin's thinking and cemented by subsequent work, we arrived at a situation where stereotyping of sex roles (Green and Madjidian, 2011) has prevented advances in theory and experimental work, as was suggested for other aspects of sexual selection theory (Ah-King et al., 2014). But maybe the field is simply lacking a better conceptual framework and adequate experimentation.

1.3 Why choose or compete?

Mate choice and competition for mates seem almost ubiquitous, but there is a cheap alternative to choosing, which is simply not to choose and to mate at random. This is, for example, widespread in marine invertebrates. One example is the palolo worm, *Palola viridis*. This polychaete releases gametes (Caspers, 1984) into the open sea and apparently the mating is random, although the release of gametes is highly coordinated and seems regulated by

the lunar cycle. One caveat here is that potentially choice may be cryptic and mate selection actually happens at the level of the gametes (Chapter 6).

The majority of metazoans, however, exercise either choice or competition for mates, or both. In other words, the costs of choice or competition must be smaller than the benefits. In addition to the easily observed costs of choice and competition, there is a large difference in the initial investment into gametes. Essentially, the argument for choice and competition is an ecological one, but it is based on the evolution of anisogamy—gametes of different sizes (Chapter 5). The traditional view holds that the sex that invests more into gametes, females, usually evolves to be the limiting resource for the other sex, males. Males, on the other hand, invest very little into the gametes and are the limited sex. In a way the two sexes we recognize use very different strategies and packaging to pass on almost equal amounts of DNA (Scharer, 2017).

Early work in sexual selection focused much on male competition (Milam, 2010), largely ignoring female choice. Even the view of aggressive males and coy females, and the associated stereotypical sex roles, are a relatively modern construct and did not really originate until after World War II (Milam, 2012). Interestingly, male mate choice saw some important early work because it was thought to be relevant in speciation, and species recognition. These studies were based on the view of the species as unit of selection and often found only weak support for the predicted male preference of their own species (Milam, 2010). As just one example, Haskins and Haskins published studies of male mate choice in guppies and some close relatives and reported evidence for male preferences for conspecific females (Haskins and Haskins, 1949, 1950). Another contemporary record of male mate choice in the context of species recognition was published by Hubbs and Delco (1960). In this paper the authors describe a conspecific preference in four species of the genus *Gambusia*, a group of livebearing fishes. They showed that most species indeed show the predicted species preference, but that *Gambusia affinis* does not.

However, the roots for modern sexual selection theory were put down by Hamilton in the 1960s when he revolutionized the way we think about the unit of selection. Until then biologists worked under the assumption that selection operates on the species, but Hamilton argued that it is the gene that is under selection (Hamilton, 1964a, 1964b), leading to a Kuhnian revolution and a complete paradigm shift (Kuhn, 1962). A precursor for Hamilton's idea was found in the works of Haldane, but Hamilton is the one who perfected the argument.

A little later, in 1972, Trivers presented his influential theory of parental investment (Trivers, 1972). He noted that the sex that invests more in their offspring usually evolves to be the choosier sex. Based on Bateman's earlier work (Bateman, 1948) on the consequences of anisogamy in *Drosophila*, he argued that females initially invest more into their gametes compared with males, and typically benefit from carefully selecting the partner they have offspring with. Females are limited by the number of eggs or reproductive events they have, whereas the male potential reproductive rate (PRR) is only limited by the number of eggs they can fertilize. Theoretical models have confirmed the role of anisogamy in the evolution of sex roles (Lehtonen et al., 2016). This makes the ecology of investment in offspring a key element in sexual selection. The majority of the research has since focused on female choice, with less focus on male choice. The above line of argument firmly established the concept of sex roles, which has been used widely, but has also been criticized as hindering research into phenomena outside of these sex roles (Hrdy, 1997; Hoquet, 2020). Typically, females can maximize their reproductive success by carefully selecting the best available male. By contrast, males are generally assumed to be far less discriminatory and should be able to maximize their reproductive success by mating with as many females as they can. This idea is captured in their often different PRR. While this view is empirically very well supported, it is also incomplete, because it does not account for the possibility of male mate choice and female competition. This has been pointed out by several authors and observations of male mate choice are now more abundant, as is theoretical literature (Krupa, 1995; Bondurianski, 2001; Servedio and Lande, 2006; Servedio, 2007; Barry and Kokko, 2010; Edward and Chapman, 2011; Fitzpatrick and Servedio, 2018) (Chapter 3).

The notion of sex roles has an additional consequence. It creates the impression of essentially binary sex roles. However, in humans we know very well that the situation is much more complex than that. In other animals we know about—for example—hermaphrodites and their sperm trading (Michiels and Bakovski, 2000), about sex-change in fishes (Ross et al., 1983), and also same-sex behavior in general (Poiani, 2010), but one wonders how much additional complexity we are missing in animal sexuality.

1.4 Mate choice

As I said above, mate choice can be viewed as a subset of sexual selection theory, which itself is a subset of natural selection as defined by Darwin, first in *On the Origin of Species* (Darwin, 1859) and in great detail in *The Descent of Man* (Darwin, 1871).

1.4.1 Female choice

It is widely accepted that females choose their mating partners, but what is the basis for that? Ultimately, the basis for our understanding of sexual selection is the assumption that ecology drives the evolution of competition and choice. There are two important facets to consider here. The argument for females being the choosier sex is based on their very large investment into gametes. Nonetheless, there are many species, including humans, in which males invest strongly into mate choice, courtship, or some aspect of raising their offspring (Buss, 2015). Such investment can level the playing field and might allow for male mate choice to evolve. This is particularly clear when considering the evolution of so-called "sex-role reversed" species, such as pipefish (Rosenqvist, 1990; Mobley et al., 2011), or tropical birds like jacanas (Emlen et al., 1989) and Nordic birds like phalaropes (Schamel et al., 2004), but it applies to all species. This said, it is also clear that females in many species continue investing into their offspring, potentially erasing all compensatory investment by males. Based on investment relative to the other sex, a broad-scale pattern of intersexual choosiness has evolved. On the other hand, males compete with other males for reproductive opportunities and, if a male increases his fitness by being more selective as compared with other males, male mate choice could evolve.

What are the main reasons for females to choose? There are direct and indirect benefits. Females— and also males as I will argue later—often choose partners that provide them with some form of a direct benefit. Or they choose males that confer an indirect, genetic benefit to their offspring.

Direct benefits are relatively easy to understand. Females select males that make a direct contribution to their offspring. This can be a space that allows females to forage or nest. It is very common for males to provide a nuptial gift to the female, like a prey item in scorpionflies (*Panorpa cognata*) (Engqvist, 2007) and other insects (Hayashi, 1998; Karlsson, 1998), or a large spermatophore, which may contain nutrients or water. This is common in many insects, including katydids. Often, the time a male is allowed to mate and transfer sperm to the female is directly correlated with the time during which the female is consuming the nuptial gift. Longer copulations lead to increased paternity (Vahed, 2006). In extreme cases the male itself becomes the nuptial gift. This has evolved several times independently and is adaptive in systems where the male has a very low probability to mate a second time. This sexual cannibalism is found in some spiders (Elgar and Schneider, 2004) and in many species of praying mantis (Barry et al., 2010). Other forms of direct benefits may include male parental care (Amundsen, 2003), provisioning of offspring (Badyaev and Ghalambor, 2001), or guarding eggs or nests (Reyer, 1984; Mol, 1996), among many expressions of paternal care. In general, females can benefit from selecting males that show good parenting ability. This is also well documented in humans, where females show preferences for mates with higher available economic resources (Mulder, 1990) and many other personality traits including ambition or stability, as well as a number of physical attributes (Buss, 2015). Not only women appreciate men that can solve problems and show cognitive ability: a similar preference was found in a small bird, the budgerigar (*Melopsittacus undulatus*) (Chen et al., 2019).

The role of indirect benefits is more difficult to study. In many cases female choice (or at least preferences) has been documented although males provide no direct benefits to the females. This is a conundrum because it is difficult to understand why females should be choosy, especially if choice is costly. Furthermore, how do male ornaments evolve under this scenario and what information do they provide for the females? Several theories have been suggested to explain this. First, ornaments could evolve to indicate good health to females and/or the ability of males to resist parasites (Hamilton and Zuk, 1982). Individual female preferences for males may aim at maximizing compatibility. This is different from a model where we assume that females have uniform preferences for more or less the same traits. Compatibility has especially been demonstrated in preferences based on the major histocompatibility complex or MHC (Wedekind, 1994; Kurtz et al., 2004). This complex of genes is important in immunodefense. Consequently, females often prefer to mate with a male that will give a female's offspring the highest diversity for the MHC (Aeschlimann et al., 2003), which plays a crucial role in immunodefense. Interestingly, this is thought to operate in humans, too (Milinski and Wedekind, 2001; Milinski, 2006). Second, they could simply indicate that the bearer of ornaments is a good survivor and may pass on viability genes to his offspring (Reynolds and Gross, 1990). Third, under models of "runaway sexual selection," male traits and female preference become genetically coupled leading to sons that show the ornament of their father and daughters that express their mother's preference (Kirkpatrick and Ryan, 1991). Finally, under "chase-away selection," females do not derive a benefit from choosing, but the male ornament exploits a pre-existing sensory bias in the females (Holland and Rice, 1998). Each of these hypotheses has some support, but it is much more limited than the evidence for direct benefits. For example, in bitterlings (*Rhodeus amarus*), a fish that lays its eggs in mussels, sperm is also regarded as an ornament, and female preferences are based on small amounts of sperm released prior to actual matings (Smith et al., 2018), suggesting that chemical communication might the mechanism for how females detect differences between males. Smith et al. (2018) also suggested that the benefits for females are indirect.

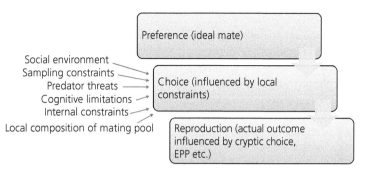

Figure 1.3 Flow chart of the process from preference to reproduction.

Furthermore, while many signals are fixed, some depend on the current condition of the signaler, reflecting current health, feeding ability, even an extended phenotype, as in, for example, the bower of bower birds (Ptilonorhynchidae) (Patricelli et al., 2006). Also, with experience they improve their skills.

How do preferences translate into actual choice and reproduction? This is not a trivial problem. Having a measurable, even consistent, preference is one thing, but does this actually translate into choice and the production of offspring? The answer is often a clear no. Mate choice is complex and complicated (Rosenthal, 2017). One can view this process as a cascade starting with a preference based on an inner template for an ideal mate (Figure 1.3). For many reasons the actual choice is much more constrained. One such reason is that the actual pool of potential mates is spatially and temporally restricted. Also, the female may not be able to wait for a better mate, and eventually has to mate based on internal constraints. This is especially important in seasonal species, like explosive breeding amphibians. Many more factors influence mate choice. For example, another extremely important factor is sexual conflict (Chapman et al., 2003).

Furthermore, mating decisions are often influenced by the social environment (Danchin et al., 2004). In almost every species, mating and all pre-mating interactions happen in a public setting, where multiple types of audiences can have an influence on mating decisions or be influenced by mating decisions (Valone, 2007). Sometimes called audience effects (Matos and Schlupp, 2004; Auld and Godin, 2015), there are a number of social

interactions that can influence mating decisions and have wide-ranging effects. One example would be mate choice copying (Agrawal, 2001; Vakirtzis, 2011; Witte et al., 2015). Mate copying occurs when females or males incorporate information about decisions made by other individuals into their own decision-making. First studied in guppies (*Poecilia reticulata*) and based on initial models by Losey et al. (Losey et al., 1986; Gibson and Höglund, 1992; Dugatkin and Godin, 1993), this phenomenon has now been discovered in many different species, such as *Drosophila*, several fishes (Schlupp et al., 1994; Applebaum and Cruz, 2000; Alonzo, 2008), birds (Galef and White, 1998), and mammals (Galef et al., 2008), including humans (Waynforth, 2007). The majority of these studies found mate copying in females, but males have also been shown to copy mate preferences (Schlupp and Ryan, 1997; Vakirtzis and Roberts, 2012). Social influences on mate choice, and mate copying in particular, have important consequences for processes like hybridization and speciation (Varela et al., 2018).

Along the lines of social influence on mate choice, deceptive behavior has also been invoked. In a study using a livebearing fish, the Atlantic molly (*Poecilia mexicana*) the authors found that males try to mislead other males to make a suboptimal mating decision (Plath et al., 2008).

Finally, not just the current social environment, but learning in general has important impacts on mate choice (Verzijden et al., 2012). Somewhat related to this, I think there will be very interesting data coming out of studies of epigenetics and sexual behavior.

Eventually, the cascade from preference via actual mating leads to actual reproduction and maternity or paternity. This final step is made complicated again because often choosers exercise cryptic choice and bias paternity or maternity. Cryptic choice in females happens via selecting sperm after the copulation has ended (Firman et al., 2017). It is also found in pipefishes, where males brood the young in pouches and exercise some control over which offspring is bred to birth via selective abortion (Paczolt and Jones, 2010) based on conditional signals reflecting female quality (Partridge et al., 2009). Cryptic choice is widespread in species with internal fertilization and allows females—at least in theory—to have social partners based on the resources or direct benefits they provide, but seek genetic paternity from males that may provide indirect, genetic benefits. Males may exercise cryptic choice by selectively allocating sperm. This has been documented for example in insects (Reinhold et al., 2002) and in fishes (Schlupp and Plath, 2005).

1.4.2 Male choice

Male choice is not simply the equivalent to female choice with the opposite sign. Female choice can lead to the evolution of traits that are detrimental to their survival, especially when males contribute little or nothing to the offspring. Because females always invest much more in the offspring via their eggs (this is how we define what a female is), it is very difficult to conceive that males would benefit from such selection on females. Males, in that sense are more dispensable, most never reproduce and die as virgins. However, the research methods and programs that have been successfully used to study female choice can also be used to understand male mate choice. Interestingly, Darwin mentioned male mate choice in *The Descent of Man* (Darwin, 1871), and thought that it was important in humans, but less so in animals (Richards, 2017) (Chapter 3).

Considering the intersexual nature of Bateman/Trivers sex roles, male choice can evolve when male investment supersedes female investment. This has happened several times, for example, in jacanas and pipefishes, but compared with mating systems where the females exercise choice this situation is

rare (Chapter 3). These mating systems are usually called sex-role reversed.

Male choosiness can also be viewed as relative to other males. If males benefit from being more selective than other males, male mate choice may evolve. Furthermore, male choice can also evolve when the benefit of choice supersedes the cost of choice and when there are differences in the quality of females (Edward and Chapman, 2011). Often this is the case when female quality differs. Then it may be adaptive for males to evolve male mate choice. In other words, male mate choice can evolve even in the absence of male investment (Schlupp, 2018), and also based on differences in sex ratios (Chapter 6).

A corollary of this is that—if males choose—they should evolve to use cues and signals to determine the quality of females, and females would benefit from evolving such traits. Hence, females may evolve ornamentation under male choice, although they are not expected to be as extreme as ornaments in males (Chapter 7). Such cues and signals could be measures of female fitness, but also ornaments and armaments that evolve in females if they compete for males (Berglund et al., 1997; Roth et al., 2014). Even in species where males contribute nothing but sperm to their offspring it can be adaptive for them to discriminate between different females and prefer certain females if the benefits from male choice outweigh the costs. Multiple mechanisms for this have been suggested (Edward and Chapman, 2011), but one easily understood pathway for how male preferences can evolve is in response to differences in female quality, especially fecundity (Chapter 3). In this case males should mate preferentially with more fecund females, or—if they mate with multiple females—mate with the most fecund female first. This idea is easily testable because in many species, including in fishes with indeterminate growth, fecundity is indeed highly correlated with body size (Dosen and Montgomerie, 2004). Patterns that are in agreement with this hypothesis have been found in many species (Chapter 4).

Rowell and Servedio (2009) discuss some of the theoretical conditions under which male choice can evolve and highlight the fundamental difference between female and male mate choice: for males under many circumstances, females are a limiting resource, but males are not a limiting resource for

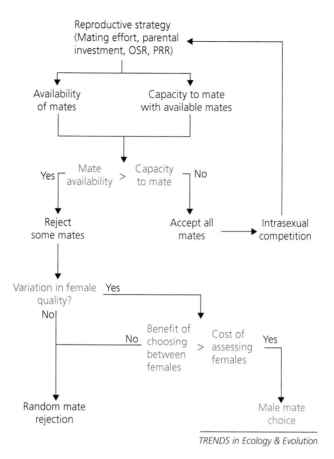

TRENDS in Ecology & Evolution

Figure 1.4 Flow chart of conditions that may lead to male mate choice. Reprinted from "The evolution and significance of male mate choice" by Dominic A. Edward and Tracey Chapman, *Trends in Ecology & Evolution*, Volume 26, Issue 12, pp. 647–654 (December 2011), DOI:10.1016 /j.tree.2011.07.012. Reprinted with permission from Elsevier.

females. This difference based on the early investment into gametes remains very important (Bonduriansky, 2001). For example, Fitzpatrick and Servedio (2017) used a population genetic model to argue that male choice will generate only weak sexual selection on females under limited circumstances, essentially saying that male mate choice seems inherently weaker as compared with female choice. In a previous paper Servedio and Lande (2006) provided detailed models for the evolution of male choice in general, also pointing out that the parameter space for the evolution of male choice is limited.

In yet another study (Barry and Kokko, 2010), the authors predict that especially with sequential mate choice (as opposed to simultaneous mate choice),

the evolution of male choice is unlikely. This sets up an interesting tension between theory and empirical studies. The review of male mate choice by Edward and Chapman (2011) proposes a simple flow chart to predict under which conditions male mate choice might evolve. The key parameters for this are a limited capacity to mate with all available females, limited mate availability, and variability in the quality of females. Finally, the benefit of mating has to exceed the cost of choice (Figure 1.4). Especially limited mate availability is an important factor that will influence male choosiness. This limitation can for example be caused by spatial or temporal shifts in the operational sex ratio (OSR). Shifts in OSR can make males the limiting sex either locally, or for a certain time period.

Finally, we have to consider that male choice may be a byproduct of female choice. Basically, the argument would be that the genetic architecture associated with choosing in females is also expressed in males. This idea, a genetic correlation and pleiotropy, is absolutely plausible (Paaby and Rockman, 2013) and should not easily be dismissed. A similar argument is debated for the expression of male-like traits or ornaments in females (Amundsen, 2000) (Chapter 7).

If both sexes in a given species choose, it is considered to be mutual mate choice (Johnstone et al., 1996; Servedio and Lande, 2006). It has been hypothesized that females and males negotiate a compromise that is acceptable to both sides (Patricelli et al., 2011). Such mutual mate choice often leads to assortative mating, where partners with trait similarity end up mating. This has been found—for example—in fishes (Myhre et al., 2012), birds (Nolan et al., 2010), and has been studied in humans (Sendova-Franks, 2013; Stulp et al., 2013) (Chapter 3). In zebra finches (*Taeniopygia guttata*), however, the notion that mutual mate choice leads to assortative mating has been challenged (Wang et al., 2017).

1.4.3 Male competition

Male competition was the first mechanism of sexual selection recognized by Darwin (1859, 1871). Males compete over females because they can benefit from mating with many females, and therefore excluding others from mating opportunities increases their own fitness. Dominant males in general seem to have more mating opportunities (Cowlishaw and Dunbar, 1991). Potentially, female choice also has a role in this, but dominance *per se* seems to be adaptive (Clutton-Brock et al., 1989). Male competition can take the form of dramatic fights between two males, sometimes to the death, but often less escalated forms of conflict can resolve the situation. On the other side of the spectrum would be sperm competition, which occurs when sperm of at least two males compete for fertilization of eggs, often inside the female's genital tract.

Especially when open fighting occurs, males often evolve formidable weapons, such as the antlers of red deer (*Cervus elaphus*) (Clutton-Brock et al., 1989), or the horns of some beetles (Emlen, 1997).

Sometimes the whole body becomes the weapon, for example when sheep ram each other head-on at full speed (Kitchener, 1988). In many cases the fights look somewhat less forceful, for example, when fishes lock their jaws and push and pull on each other (Bischof, 1996). Often, when weapons have evolved, protective structures also evolve to mitigate the potential damage inflicted by opponents. One example for that would be the mane of male lions (West and Packer, 2002). This can also be used to illustrate the dual function that weapons and protective structures can have: they may also be sexual signals and ornaments that females respond to. Male competition is often a very visible and dramatic behavior and one has to wonder if that has biased studies and reporting of the phenomenon. Female competition seems to be subtler and may have been overlooked more often (Rosvall, 2011).

Very often, large numbers of males would have no chance in an open fight. Winning a conflict is mostly determined by size, as demonstrated for the green swordtail, *Xiphophorus hellerii* (Franck and Ribowski, 1989), and smaller males almost invariably lose. In some species males wait and grow before they challenge another male, but often alternative mating strategies have evolved (Gross, 1996; Oliveira et al., 2008). In these cases, some males try to undermine the mating investment of other males in various ways. They can mimic females and release sperm instead of eggs in the nest of another male (Gonçalves et al., 1996). Or they can try to sneak copulations with a female, avoiding the costs associated with courtship (Ryan and Causey, 1989). These sneakers can be distinct morphs, but in some species males can switch between roles (Kodric, 1986; Travis and Woodward, 1989). Finally, some males may be satellite males that try to intercept females as they are trying to approach other, courting males (Arak, 1988). This is very common in lek breeding systems, where courters gather and display to choosers. Often this is done in the same place every year.

Males may compete for direct access to females, but they can also compete over resources that will subsequently attract females. This is very prominent in many species of birds where males compete over breeding territories of varying quality, which are in turn evaluated by females.

Sperm competition shows that female choice and male competition can intersect. While male competition can be used to explain why it may be beneficial for males to inseminate females multiple times or with large amounts of semen as a response to the possibility of other males also inseminating the same female, the female may use cryptic female choice to select sperm from a certain male. Because armaments can also be ornaments and traits that are adaptive in fights may also be attractive to females, so female choice and male competition can be intimately intertwined (Berglund et al., 1996).

1.4.4 Female competition

Females often compete with each other over resources, although this can be less obvious as compared with males. Openly aggressive encounters between females are rare, but female competition is probably widespread. However, documentation of female competition over males seems to be relatively rare (Rosvall, 2013; Cain and Rosvall, 2014). There are, however, some studies like the one by Petrie (1983) that clearly shows female competition for males in a lek situation, concluding that "females compete for small fat males." Most female competition, however, seems to be about resources rather than directly for males (Scharnweber et al., 2011a, 2011b). A field study in a livebearing fish, the Atlantic molly (Heubel and Plath, 2008), pointed toward intensive species competition over males and other resources. This view is supported by recent experimental work on female aggression and competition (Makowicz and Schlupp, 2013, 2015; Makowicz et al., 2016, 2018, 2020) in Amazon mollies (*Poecilia formosa*), another livebearing fish. This suggests that we need more work on within-species competition, as we seem to know relatively little about within-species female competition.

A string of papers has raised important questions about female–female competition and its evolutionary consequences. One especially important aspect here is the distinction between the roles of social selection (Tobias et al., 2012) and sexual selection (Cain and Rosvall, 2014) (Chapter 8). Another important aspect is the potential evolution of armaments and weapons in females. We have some knowledge of this in sex-role reversed species (Watson and Simmons, 2010), but little beyond.

1.5 Definitions

Here I want to provide an operational definition of what "mate choice" is. I am making reference to the definition recently suggested by Rosenthal (2017) and also used by Schlupp (2018): "Mate choice can be defined as any aspect of an animal's phenotype that leads to it being more likely to engage in sexual activity with certain individuals than with others." Note that this definition parts dramatically from the problematic traditional usage of sex roles (Ah-King, 2013b). Consequently, Rosenthal (2017) replaces female and male with the terms chooser and courter, which can be of any sex. I agree with this definition, but in this book, I want to retain the usage of male and female as a heuristic tool, to reflect the existing difference in the ecology of early investment into gametes, without acknowledging specific sex roles. Furthermore, I want to emphasize that in my view both sexes can be chooser and courters at the same time. I think that we eventually have to realize that mate choice is best understood as a continuum, with the traditional sex roles of male and female confined to the extreme ends. I propose that in reality in most mating systems, females and males both have preferences, exercise choice, and resolve the underlying sexual conflict with some form of reciprocal mate choice.

Similarly, the term "preference" needs to be defined. Again, I use Rosenthal (2017): "a chooser's internal representation of courter traits that predisposes it to mate with some phenotypes over others." The difference between choice and preference is that we can assess choice by measuring actual sexual behaviors and outcomes, while preferences can also be measured indirectly, for example, by using association times or preference functions (Wagner, 1998).

An ornament is a trait (or a combination of traits) that is likely to have arisen via sexual selection and plays a role in mate choice by making the bearer attractive to choosers, often at a cost to survival. Ornaments are often sexually dimorphic, but they do not have to be.

Finally, the terms male and female competition also need to be defined. Broadly speaking these are phenomena where individuals of the same sex compete for access to members of the other sex. This is most obvious in species where individuals of one sex actually fight with another, but the interactions can be much more subtle.

Both within-sex competition and between-sex choice lead to differential reproduction, which is then exposed to selection.

1.6 Short summary

This chapter is introducing the key topics of the book—male mate choice and female competition—relative to each other, and relative to the very established mechanisms of female choice and male competition. It is important to realize that although male mate choice is common and likely more important than currently realized, it is unlikely to have the same effects on females that female choice has on males.

1.7 Recommended reading

The book by Rosenthal (2017) on mate choice is an excellent and comprehensive account of mate choice.

The short review by Edward and Chapman (2011) captures a lot of the thinking on male mate choice.

Male mate choice in insects is covered by Bonduriansky (2001).

Important historical background is provided by Milam (2010) and Richards (2017).

1.8 Bibliographic information

In February 2018 I conducted a search in the bibliographic database Web of Science to find out how many peer-reviewed scientific papers were published in the areas that are the topic of this book. I first searched very broadly for sexual selection, and then conducted a narrow search for the other terms, male and female choice, as well as female competition and male competition. In Web of Science one has many different options to filter results, but what I did with my search for sexual selection was very simple: I used the core database and typed the terms "Sexual" and "Selection" in the search window in the rubric "Topic." The time period covered was from 1900 to 2017. This search will find papers in the database that contain the words "Sexual" *and* "Selection." This simple search found roughly 27,000 articles. The graph in Figure 1.5 indicates a low number of papers per year until an explosive growth occurs in the field around 1990. This seems to coincide with the publication of Andersson's formative book (Andersson, 1994), but it is hard to pinpoint a single event that might explain this pattern.

I conducted two much more narrow searches on the core terms relevant to this book. First, I studied female choice. This time I used all available databases in Web of Science but restricted the search to papers

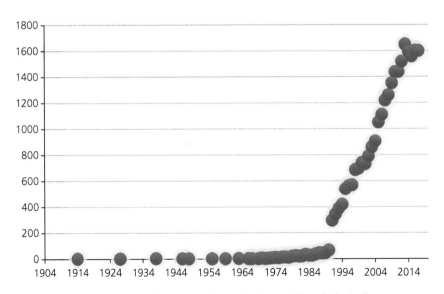

Figure 1.5 Number of papers published per year that contain the words "Sexual Selection."

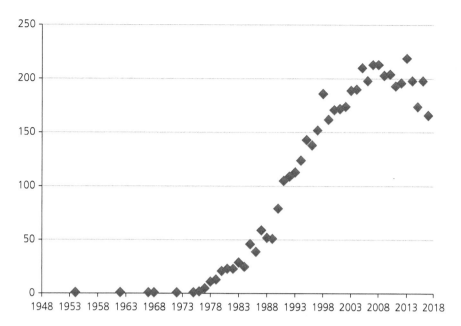

Figure 1.6 Number of publications per year for the term "Female Choice."

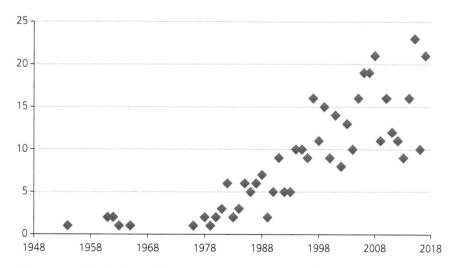

Figure 1.7 Number of publications for the term "Male Choice."

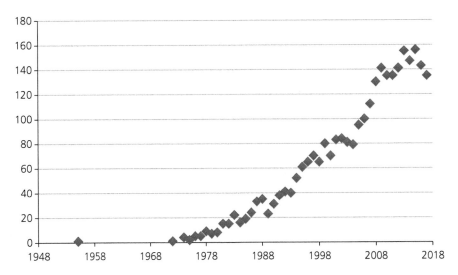

Figure 1.8 Number of publications for the term "Male Competition."

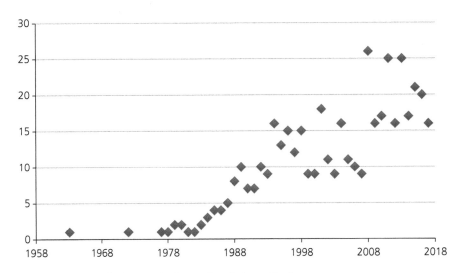

Figure 1.9 Number of publications for the term "Female Competition."

that had the terms "Female" and "Choice" next to each other. This ignores some papers but should give an overview of studies that were truly addressing female choice. This search yielded 5073 hits (Figure 1.6). A similar search for "Male Choice" yielded 401 records (Figure 1.7), not even 10 percent of the papers on female mate choice. Using a similar method, I extended this search to "Male Competition" (2826 records, Figure 1.8) and "Female Competition" (440 records, Figure 1.9). All graphs show an increase in number of studies published per year, which is in agreement with the overall increase in studies in sexual selection. However, clearly, there is a lot less work done on male choice and female competition. As a consequence of the much lower absolute numbers the graphs for male choice and female competition seem to show less clear trends. It also seems as if the numbers of studies on male competition and female choice are reaching a plateau.

1.9 References

AESCHLIMANN, P. B., HABERLI, M. A., REUSCH, T. B. H., BOEHM, T., & MILINSKI, M. 2003. Female sticklebacks *Gasterosteus aculeatus* use self-reference to optimize MHC allele number during mate selection. *Behavioral Ecology and Sociobiology,* 54, 119–26.

AGRAWAL, A. F. 2001. The evolutionary consequences of mate copying on male traits. *Behavioral Ecology & Sociobiology,* 51, 33–40.

AH-KING, M. 2013a. *Challenging Popular Myths of Sex, Gender and Biology.* Cham, Switzerland: Springer.

AH-KING, M. 2013b. On anisogamy and the evolution of "sex roles." *Trends in Ecology & Evolution,* 28, 1–2.

AH-KING, M., BARRON, A. B., & HERBERSTEIN, M. E. 2014. Genital evolution: why are females still understudied? *PLoS Biology,* 12. https://doi.org/10.1371/journal.pbio.1001851.

AH-KING, M. & GOWATY, P. A. 2016. A conceptual review of mate choice: stochastic demography, within-sex phenotypic plasticity, and individual flexibility. *Ecology and Evolution,* 6, 4607–42.

ALONZO, S. H. 2008. Female mate choice copying affects sexual selection in wild populations of the ocellated wrasse. *Animal Behaviour,* 75, 1715–23.

ALONZO, S. H. & SERVEDIO, M. R. 2019. Grey zones of sexual selection: why is finding a modern definition so hard? *Proceedings of the Royal Society of London—Series B: Biological Sciences,* 286, 20191325.

AMUNDSEN, T. 2000. Why are female birds ornamented? *Trends in Ecology & Evolution,* 15, 149–55.

AMUNDSEN, T. 2003. Fishes as models in studies of sexual selection and parental care. *Journal of Fish Biology,* 63, 17–52.

ANDERSSON, M. 1994. *Sexual Selection.* Princeton, NJ: Princeton University Press.

APPLEBAUM, S. L. & CRUZ, A. 2000. The role of mate-choice copying and disruption effects in mate preference determination of *Limia perugiae* (Cyprinodontiformes, Poeciliidae). *Ethology,* 106, 933–44.

ARAK, A. 1988. Callers and satellites in the natterjack toad: evolutionarily stable decision rules. *Animal Behaviour,* 36, 416–32.

AULD, H. L. & GODIN, J. G. J. 2015. Sexual voyeurs and copiers: social copying and the audience effect on male mate choice in the guppy. *Behavioral Ecology and Sociobiology,* 69, 1795–807.

BADYAEV, A. V. & GHALAMBOR, C. K. 2001. Evolution of life histories along elevational gradients: trade-off between parental care and fecundity. *Ecology,* 82, 2948–60.

BARRY, K. L., HOLWELL, G. I., & HERBERSTEIN, M. E. 2010. Multimodal mate assessment by male praying mantids in a sexually cannibalistic mating system. *Animal Behaviour,* 79, 1165–72.

BARRY, K. L. & KOKKO, H. 2010. Male mate choice: why sequential choice can make its evolution difficult. *Animal Behaviour,* 80, 163–9.

BATEMAN, A. J. 1948. Intra-sexual selection in *Drosophila. Heredity,* 2, 349–68.

BERGLUND, A., BISAZZA, A., & PILASTRO, A. 1996. Armaments and ornaments: an evolutionary explanation of traits of dual utility. *Biological Journal of the Linnean Society,* 58, 385–99.

BERGLUND, A. & ROSENQVIST, G. 2003. Sex role reversal in pipefish. *In:* SLATER, P. J. B., ROSENBLATT, J. S., SNOWDON, C. T., & ROPER, T. J. (eds.) *Advances in the Study of Behavior, Vol 32.* Cambridge, MA: Academic Press, pp. 131–67.

BERGLUND, A., ROSENQVIST, G., & BERNET, P. 1997. Ornamentation predicts reproductive success in female pipefish. *Behavioral Ecology and Sociobiology,* 40, 145–50.

BISCHOF, C. 1996. Diversity in agonistic behavior of croaking gouramis (*Trichopsis vittata, T. schalleri,* and *T. pumila*; Anabantoidei) and the paradise fish (*Macropodus opercularis*; Anabantoidei). *Aggressive Behavior,* 22, 447–55.

BONDURIANSKY, R. 2001. The evolution of male mate choice in insects: a synthesis of ideas and evidence. *Biological Reviews,* 76, 305–39.

BORODITSKY, L. 2001. Does language shape thought?: Mandarin and English speakers' conceptions of time. *Cognitive Psychology,* 43, 1–22.

BUSS, D. 2015. *Evolutionary Psychology.* Abingdon, UK: Routledge.

CAIN, K. E. & ROSVALL, K. A. 2014. Next steps for understanding the selective relevance of female-female competition. *Frontiers in Ecology and Evolution,* 2. https://doi.org/10.3389/fevo.2014.00032.

CANDOLIN, U. 2003. The use of multiple cues in mate choice. *Biological Reviews,* 78, 575–95.

CASPERS, H. 1984. Spawning periodicity and habitat of the palolo worm *Eunice viridis* (Polychaeta: Eunicidae) in the Samoan Islands. *Marine Biology,* 79, 229–36.

CHAPMAN, T., ARNQVIST, G., BANGHAM, J., & ROWE, L. 2003. Sexual conflict. *Trends in Ecology & Evolution,* 18, 41–7.

CHEN, J., ZOU, Y., SUN, Y.-H., & TEN CATE, C. 2019. Problem-solving males become more attractive to female budgerigars. *Science,* 363, 166–7.

CLUTTON-BROCK, T. H., HIRAIWA-HASEGAWA, M., & ROBERTSON, A. 1989. Mate choice on fallow deer leks. *Nature,* 340, 463–5.

COWLISHAW, G. & DUNBAR, R. I. M. 1991. Dominance rank and mating success in male primates. *Animal Behaviour,* 41, 1045–56.

DANCHIN, E., GIRALDEAU, L. A., VALONE, T. J., & WAGNER, R. H. 2004. Public information: from nosy neighbors to cultural evolution. *Science*, 305, 487–91.

DARGENT, F., CHEN, L., FUSSMANN, G. F., GHALAMBOR, C. K. & HENDRY, A. P. 2019. Female preference for novel males constrains the contemporary evolution of assortative mating in guppies. *Behavioral Ecology*, 30, 646–57.

DARWIN, C. 1859. *On the Origin of Species.* London: John Murray.

DARWIN, C. 1871. *The Descent of Man, and Selection in Relation to Sex.* London: John Murray.

DOSEN, L. D. & MONTGOMERIE, R. 2004. Female size influences mate preferences of male guppies. *Ethology*, 110, 245–55.

DUGATKIN, L. A. & GODIN, J.-G. J. 1993. Female mate copying in the guppy (*Poecilia reticulata*): age-dependent effects. *Behavioral Ecology*, 4, 289–92.

DUNSWORTH, H. M. 2020. Expanding the evolutionary explanations for sex differences in the human skeleton. *Evolutionary Anthropology: Issues, News, and Reviews*, 29, 108–16.

EDWARD, D. A. & CHAPMAN, T. 2011. The evolution and significance of male mate choice. *Trends in Ecology & Evolution*, 26, 647–54.

ELGAR, M. A. & SCHNEIDER, J. M. 2004. Evolutionary significance of sexual cannibalism. *Advances in the Study of Behavior*, 34, 135–63.

EMLEN, D. J. 1997. Alternative reproductive tactics and male-dimorphism in the horned beetle *Onthophagus acuminatus* (Coleoptera: Scarabaeidae). *Behavioral Ecology and Sociobiology*, 41, 335–41.

EMLEN, S. T., DEMONG, N. J., & EMLEN, D. J. 1989. Experimental induction of infanticide in female wattled jacanas. *Auk*, 106, 1–7.

ENGQVIST, L. 2007. Sex, food and conflicts: nutrition dependent nuptial feeding and pre-mating struggles in scorpionflies. *Behavioral Ecology and Sociobiology*, 61, 703–10.

FIRMAN, R. C., GASPARINI, C., MANIER, M. K., & PIZZARI, T. 2017. Postmating female control: 20 years of cryptic female choice. *Trends in Ecology & Evolution*, 32, 368–82.

FITZPATRICK, C. L. & SERVEDIO, M. R. 2017. Male mate choice, male quality, and the potential for sexual selection on female traits under polygyny. *Evolution*, 71, 174–83.

FITZPATRICK, C. L. & SERVEDIO, M. R. 2018. The evolution of male mate choice and female ornamentation: a review of mathematical models. *Current Zoology*, 64, 323–33.

FRANCK, D. & RIBOWSKI, A. 1989. Escalating fights for rank-order position between male swordtails (*Xiphophorus helleri*): effects of prior rank-order experience and information transfer. *Behavioral Ecology & Sociobiology*, 24, 133–44.

GALEF, B. G., LIM, T. C. W., & GILBERT, G. S. 2008. Evidence of mate choice copying in Norway rats, *Rattus norvegicus*. *Animal Behaviour*, 75, 1117–23.

GALEF, B. G. & WHITE, D. J. 1998. Mate-choice copying in Japanese quail, *Coturnix coturnix japonica*. *Animal Behaviour*, 55, 545–52.

GIBSON, R. M., BRADBURY, J. W., & VEHRENCAMP, S. L. 1991. Mate choice in lekking sage grouse revisited the roles of vocal display female site fidelity and copying. *Behavioral Ecology*, 2, 165–80.

GIBSON, R. M. & HÖGLUND, J. 1992. Copying and sexual selection. *Trends in Ecology & Evolution*, 7, 229–32.

GONÇALVES, E. J., ALMADA, V. C., OLIVEIRA, R. F., & SANTOS, A. J. 1996. Female mimicry as a mating tactic in males of the blenniid fish *Salaria pavo*. *Journal of the Marine Biological Association of the United Kingdom*, 76, 529–38.

GREEN, K. K. & MADJIDIAN, J. A. 2011. Active males, reactive females: stereotypic sex roles in sexual conflict research? *Animal Behaviour*, 81, 901–7.

GROSS, M. R. 1996. Alternative reproductive strategies and tactics: diversity within sexes. *Trends in Ecology & Evolution*, 11, 92–8.

HAMILTON, W. D. 1964a. The genetical evolution of social behaviour I. *Journal of Theoretical Biology*, 7, 1–16.

HAMILTON, W. D. 1964b. The genetical evolution of social behaviour II. *Journal of Theoretical Biology*, 7, 17–52.

HAMILTON, W. D. & ZUK, M. 1982. Heritable true fitness and bright birds: a role for parasites. *Science*, 218, 384–7.

HASKINS, C. P. & HASKINS, E. F. 1949. The role of sexual selection as an isolating mechanism in three species of Poeciliid fishes. *Evolution*, 3, 160–9.

HASKINS, C. P. & HASKINS, E. F. 1950. Factors governing sexual selection as an isolating mechanism in the Poeciliid fish *Lebistes reticulatus*. *PNAS*, 36, 464–76.

HAYASHI, F. 1998. Multiple mating and lifetime reproductive output in female dobsonflies that receive nuptial gifts. *Ecological Research*, 13, 283–9.

HEUBEL, K. U. & PLATH, M. 2008. Influence of male harassment and female competition on female feeding behaviour in a sexual-asexual mating complex of mollies (*Poecilia mexicana*, *P. formosa*). *Behavioral Ecology and Sociobiology*, 62, 1689–99.

HIRAIWA-HASEGAWA, M. 2000. The sight of the peacock's tail makes me sick: the early arguments on sexual selection. *Journal of Biosciences*, 25, 11–18.

HO, S. S., BROSSARD, D. & SCHEUFELE, D. A. 2008. Effects of value predispositions, mass media use, and knowledge on public attitudes toward embryonic stem cell research. *International Journal of Public Opinion Research*, 20, 171–92.

HOLLAND, B. & RICE, W. R. 1998. Perspective: Chase-away sexual selection: antagonistic seduction versus resistance. *Evolution*, 52, 1–7.

HOQUET, T. 2020. Bateman (1948): rise and fall of a paradigm? *Animal Behaviour*, 164, 223–31

HRDY, S. B. 1997. Raising Darwin's consciousness—female sexuality and the prehominid origins of patriarchy. *Human Nature—an Interdisciplinary Biosocial Perspective*, 8, 1–49.

HUBBS, C. & DELCO, E. A. 1960. Mate preference in males of four species of Gambusiine fishes. *Evolution*, 14, 145–52.

JAMIESON, C. 2013. Gun violence research: history of the federal funding freeze. Newtown tragedy may lead to lifting of freeze in place since 1996. *Psychological Science Agenda*. https://www.apa.org/science/about/psa/2013/02/gun-violence (accessed November 16, 2020).

JOHNSTONE, R. A., REYNOLDS, J. D., & DEUTSCH, J. C. 1996. Mutual mate choice and sex differences in choosiness. *Evolution*, 50, 1382–91.

KARLSSON, B. 1998. Nuptial gifts, resource budgets, and reproductive output in a polyandrous butterfly. *Ecology*, 79, 2931–40.

KELLERMANN, A. L., RIVARA, F. P., RUSHFORTH, N. B., BANTON, J. G., REAY, D. T., FRANCISCO, J. T., LOCCI, A. B., PRODZINSKI, J., HACKMAN, B. B., & SOMES, G. 1993. Gun ownership as a risk factor for homicide in the home. *New England Journal of Medicine*, 329, 1084–91.

KIRKPATRICK, M. & RYAN, M. J. 1991. The evolution of mating preferences and the paradox of the lek. *Nature*, 350, 33–8.

KITCHENER, A. 1988. An analysis of the forces of fighting of the blackbuck (*Antilope cervicapra*) and the bighorn sheep (*Ovis canadensis*) and the mechanical design of the horn of bovids. *Journal of Zoology*, 214, 1–20.

KODRIC, B. A. 1986. Satellites and sneakers opportunistic male breeding tactics in pupfish (*Cyprinodon pecosensis*). *Behavioral Ecology & Sociobiology*, 19, 425–32.

KRUPA, J. J. 1995. how likely is male mate choice among anurans. *Behaviour*, 132, 643–64.

KUHN, T. S. 1962. *The Structure of Scientific Revolutions*. Chicago, IL: University of Chicago Press.

KURTZ, J., KALBE, M., AESCHLIMANN, P. B., HAEBERLI, M. A., WEGNER, K. M., REUSCH, T. B. H., & MILINSKI, M. 2004. Major histocompatibility complex diversity influences parasite resistance and innate immunity in sticklebacks. *Proceedings of the Royal Society of London—Series B: Biological Sciences*, 271, 197–204.

LEBOEUF, B. J. 1972. Sexual behavior in the northern elephant seal *Mirounga Angustirostris*. *Behaviour*, 41, 1–26.

LEHTONEN, J., PARKER, G. A., & SCHÄRER, L. 2016. Why anisogamy drives ancestral sex roles. *Evolution*, 70, 1129–35.

LINDSAY, W. R., ANDERSSON, S., BERERHI, B., HÖGLUND, J., JOHNSEN, A., KVARNEMO, C., LEDER, E. H., LIFJELD, J. T., NINNES, C. E., OLSSON, M., PARKER, G. A., PIZZARI, T., QVARNSTRÖM, A., SAFRAN, R. J., SVENSSON, O., & EDWARDS, S. V. 2019. Endless forms of sexual selection. *PeerJ*, 7, e7988.

LIU, Z., CAI, Y., WANG, Y., NIE, Y., ZHANG, C., XU, Y., ZHANG, X., LU, Y., WANG, Z., POO, M., & SUN, Q. 2018. Cloning of macaque monkeys by somatic cell nuclear transfer. *Cell*, 172, 881–7.

LOSEY, G. S., STANTON, F. G., TELECKY, T. M., TYLER III, W. A., & ZOOLOGY 691 GRADUATE SEMINAR CLASS. 1986. Copying others, an evolutionarily stable strategy for mate choice: a model. *American Naturalist*, 128, 653–64.

LÜPOLD, S., MANIER, M. K., PUNIAMOORTHY, N., SCHOFF, C., STARMER, W. T., LUEPOLD, S. H. B., BELOTE, J. M., & PITNICK, S. 2016. How sexual selection can drive the evolution of costly sperm ornamentation. *Nature*, 533, 535–38.

MAKOWICZ, A. M., MOORE, T., & SCHLUPP, I. 2018. Clonal fish are more aggressive to distant relatives in a low resource environment. *Behaviour*, 155, 351–67.

MAKOWICZ, A. M., MURRAY, L., & SCHLUPP, I. 2020. Size, species and audience type influence heterospecific female-female competition. *Animal Behaviour*, 159, 47–58.

MAKOWICZ, A. M. & SCHLUPP, I. 2013. The direct costs of living in a sexually harassing environment. *Animal Behaviour*, 85, 569–77.

MAKOWICZ, A. M. & SCHLUPP, I. 2015. Effects of female-female aggression in a sexual/unisexual species complex. *Ethology*, 121, 903–14.

MAKOWICZ, A. M., TIEDEMANN, R., STEELE, R. N., & SCHLUPP, I. 2016. Kin recognition in a clonal fish, *Poecilia formosa*. *PLoS ONE*, 11. https://doi.org/10.1371/journal.pone.0158442.

MATOS, R. & SCHLUPP, I. 2004. Performing in front of an audience—signalers and the social environment. *In*: MCGREGOR, P. K. (ed.) *Animal Communication Networks*. Cambridge: Cambridge University Press, pp. 63–83.

MCLAREN, A. 2001. Ethical and social considerations of stem cell research. *Nature*, 414, 129.

MICHIELS, N. K. & BAKOVSKI, B. 2000. Sperm trading in a hermaphroditic flatworm: reluctant fathers and sexy mothers. *Animal Behaviour*, 59, 319–25.

MILAM, E. L. 2010. *Looking For a Few Good Males*. Baltimore, MD: The Johns Hopkins University Press.

MILAM, E. L. 2012. Making males aggressive and females coy: gender across the animal-human boundary. *Signs,* 37, 935–59.

MILINSKI, M. 2006. The major histocompatibility complex, sexual selection, and mate choice. *Annual Review of Ecology Evolution and Systematics,* 37, 159–86.

MILINSKI, M. & WEDEKIND, C. 2001. Evidence for MHC-correlated perfume preferences in humans. *Behavioral Ecology,* 12, 140–9.

MOBLEY, K. B., KVARNEMO, C., AHNESJÖ, I., PARTRIDGE, C., BERGLUND, A., & JONES, A. G. 2011. The effect of maternal body size on embryo survivorship in the broods of pregnant male pipefish. *Behavioral Ecology and Sociobiology,* 65, 1169–77.

MOL, J. H. 1996. Impact of predation on early stages of the armoured catfish *Hoplosternum thoracatum* (Siluriformes-Callichthyidae) and implications for the syntopic occurrence with other related catfishes in a neotropical multi-predator swamp. *Oecologia,* 107, 395–410.

MULDER, M. B. 1990. Kipsigis womens preferences for wealthy men—evidence for female choice in mammals. *Behavioral Ecology and Sociobiology,* 27, 255–64.

MYHRE, L. C., DE JONG, K., FORSGREN, E., & AMUNDSEN, T. 2012. Sex roles and mutual mate choice matter during mate sampling. *American Naturalist,* 179, 741–55.

NISBET, M. C. 2005. The competition for worldviews: values, information, and public support for stem cell research. *International Journal of Public Opinion Research,* 17, 90–112.

NOLAN, P. M., DOBSON, F. S., NICOLAUS, M., KARELS, T. J., MCGRAW, K. J., & JOUVENTIN, P. 2010. Mutual mate choice for colorful traits in king penguins. *Ethology,* 116, 635–44.

OLIVEIRA, R. F., TABORSKY, M., & BROCKMANN, H. J. (eds.) 2008. *Alternative Reproductive Tactics.* Cambridge: Cambridge University Press.

PAABY, A. B. & ROCKMAN, M. V. 2013. The many faces of pleiotropy. *Trends in Genetics,* 29, 66–73.

PACZOLT, K. A. & JONES, A. G. 2010. Post-copulatory sexual selection and sexual conflict in the evolution of male pregnancy. *Nature,* 464, 401.

PARTAN, S. R. & MARLER, P. 2005. Issues in the classification of multimodal communication signals. *American Naturalist,* 166, 231–45.

PARTRIDGE, C., AHNESJÖ, I., KVARNEMO, C., MOBLEY, K. B., BERGLUND, A., & JONES, A. G. 2009. The effect of perceived female parasite load on postcopulatory male choice in a sex-role-reversed pipefish. *Behavioral Ecology and Sociobiology,* 63, 345–54.

PATRICELLI, G. L., COLEMAN, S. W., & BORGIA, G. 2006. Male satin bowerbirds, *Ptilonorhynchus violaceus,* adjust their display intensity in response to female startling: an experiment with robotic females. *Animal Behaviour,* 71, 49–59.

PATRICELLI, G. L., KRAKAUER, A. H., & MCELREATH, R. 2011. Assets and tactics in a mating market: economic models of negotiation offer insights into animal courtship dynamics on the lek. *Current Zoology,* 57, 225–36.

PETRIE, M. 1983. Female moorhens compete for small fat males. *Science,* 220, 413–15.

PETRIE, M. 1994. Improved growth and survival of offspring of peacocks with more elaborate trains. *Nature,* 371, 598–9.

PLATH, M., RICHTER, S., TIEDEMANN, R., & SCHLUPP, I. 2008. Male fish deceive competitors about mating preferences. *Current Biology,* 18, 1138–41.

POIANI, A. 2010. *Animal Homosexuality.* Cambridge: Cambridge University Press.

REINHOLD, K., KURTZ, J., & ENGQVIST, L. 2002. Cryptic male choice: sperm allocation strategies when female quality varies. *Journal of Evolutionary Biology,* 15, 201–9.

REYER, H. U. 1984. Investment and relatedness a cost-benefit analysis of breeding and helping in the pied kingfisher *Ceryle rudis. Animal Behaviour,* 32, 1163–78.

REYNOLDS, J. D. & GROSS, M. R. 1990. Costs and benefits of female mate choice: is there a lek paradox? *The American Naturalist,* 136, 230–43.

RICHARDS, E. 2017. *Darwin and the Making of Sexual Selection.* Chicago, IL: The University of Chicago Press.

ROSENQVIST, G. 1990. Male mate choice and female female competition for mates in the pipefish *Nerophis ophidion. Animal Behaviour,* 39, 1110–15.

ROSENTHAL, G. G. 2017. *Mate Choice.* Princeton, NJ: Princeton University Press.

ROSS, R. M., LOSEY, G. S., & DIAMOND, M. 1983. Sex change in a coral reef fish dependence of stimulation and inhibition on relative size. *Science,* 221(4610), 574–5.

ROSVALL, K. A. 2011. Intrasexual competition in females: evidence for sexual selection? *Behavioral Ecology,* 22, 1131–40.

ROSVALL, K. A. 2013. Proximate perspectives on the evolution of female aggression: good for the gander, good for the goose? *Philosophical Transactions of the Royal Society B—Biological Sciences,* 368, 20130083.

ROTH, O., SUNDIN, J., BERGLUND, A., ROSENQVIST, G., & WEGNER, K. M. 2014. Male mate choice relies on major histocompatibility complex class I in a sex-role-reversed pipefish. *Journal of Evolutionary Biology,* 27, 929–38.

ROWELL, J. & SERVEDIO, M. 2009. Gentlemen prefer blondes: the evolution of mate preference among strategically allocated males. *The American Naturalist,* 173, 12–25.

RYAN, M. J. & CAUSEY, B. A. 1989. Alternative mating behavior in the swordtails *Xiphophorus nigrensis* and *Xiphophorus pygmaeus* (Pisces, Poeciliidae). *Behavioral Ecology & Sociobiology*, 24, 341–8.

SAYADI, A., BARRIO, A. M., IMMONEN, E., DAINAT, J., BERGER, D., TELLGREN-ROTH, C., NYSTEDT, B., & ARNQVIST, G. 2019. The genomic footprint of sexual conflict. *Nature Ecology & Evolution*, 3, 1725–30.

SCHAMEL, D., TRACY, D. M., & LANK, D. B. 2004. Male mate choice, male availability and egg production as limitations on polyandry in the red-necked phalarope. *Animal Behaviour*, 67, 847–53.

SCHARER, L. 2017. The varied ways of being male and female. *Molecular Reproduction and Development*, 84, 94–104.

SCHARNWEBER, K., PLATH, M., & TOBLER, M. 2011a. Feeding efficiency and food competition in coexisting sexual and asexual livebearing fishes of the genus *Poecilia*. *Environmental Biology of Fishes*, 90, 197–205.

SCHARNWEBER, K., PLATH, M., WINEMILLER, K. O., & TOBLER, M. 2011b. Dietary niche overlap in sympatric asexual and sexual livebearing fishes *Poecilia* spp. *Journal of Fish Biology*, 79, 1760–73.

SCHLUPP, I. 2018. Male mate choice in livebearing fishes: an overview. *Current Zoology*, 64, 393–403.

SCHLUPP, I., MARLER, C., & RYAN, M. J. 1994. Benefit to male sailfin mollies of mating with heterospecific females. *Science*, 263, 373–4.

SCHLUPP, I. & PLATH, M. 2005. Male mate choice and sperm allocation in a sexual/asexual mating complex of *Poecilia* (Poeciliidae, Teleostei). *Biology Letters*, 1, 169–71.

SCHLUPP, I. & RYAN, M. J. 1997. Male sailfin mollies (*Poecilia latipinna*) copy the mate choice of other males. *Behavioral Ecology*, 8, 104–7.

SENDOVA-FRANKS, A. 2013. Human mutual mate choice for height results in a compromise. *Animal Behaviour*, 86, 2–2.

SERVEDIO, M. R. 2007. Male versus female mate choice: sexual selection and the evolution of species recognition via reinforcement. *Evolution*, 61, 2772–89.

SERVEDIO, M. R. & LANDE, R. 2006. Population genetic models of male and mutual mate choice. *Evolution*, 60, 674–85.

SMITH, C., SPENCE, R., & REICHARD, M. 2018. Sperm is a sexual ornament in rose bitterling. *Journal of Evolutionary Biology*, 31, 1610–22.

STULP, G., BUUNK, A. P., POLLET, T. V., NETTLE, D., & VERHULST, S. 2013. Are human mating preferences with respect to height reflected in actual pairings? *PLoS ONE*, 8, e54186.

TOBIAS, J. A., MONTGOMERIE, R., & LYON, B. E. 2012. The evolution of female ornaments and weaponry: social selection, sexual selection and ecological competi-tion. *Philosophical Transactions of the Royal Society B—Biological Sciences*, 367, 2274–93.

TRAVIS, J. & WOODWARD, B. D. 1989. Social context and courtship flexibility in male sailfin mollies *Poecilia latipinna* (Pisces: Poeciliidae). *Animal Behaviour*, 38, 1001–11.

TRIVERS, R. (ed.) 1972. *Parental Investment and Sexual Selection*. Chicago, IL: Aldine.

VAHED, K. 2006. Larger ejaculate volumes are associated with a lower degree of polyandry across bushcricket taxa. *Proceedings of the Royal Society of London—Series B: Biological Sciences*, 273, 2387–94.

VAKIRTZIS, A. 2011. Mate choice copying and noninde-pendent mate choice: a critical review. *Annales Zoologici Fennici*, 48, 91–107.

VAKIRTZIS, A. & ROBERTS, S. C. 2012. Human noninde-pendent mate choice: is model female attractiveness everything? *Evolutionary Psychology*, 10, 225–37.

VALONE, T. J. 2007. From eavesdropping on performance to copying the behavior of others: a review of public information use. *Behavioral Ecology and Sociobiology*, 62, 1–14.

VARELA, S. A. M., MATOS, M., & SCHLUPP, I. 2018. The role of mate-choice copying in speciation and hybridization. *Biological Reviews*, 93, 1304–22.

VERZIJDEN, M. N., TEN CATE, C., SERVEDIO, M. R., KOZAK, G. M., BOUGHMAN, J. W., & SVENSSON, E. I. 2012. The impact of learning on sexual selection and speciation. *Trends in Ecology & Evolution*, 27, 511–19.

WAGNER, W. E. J. 1998. Measuring female mating prefer-ences. *Animal Behaviour*, 55, 1029–42.

WANG, D. P., FORSTMEIER, W., & KEMPENAERS, B. 2017. No mutual mate choice for quality in zebra finches: time to question a widely held assumption. *Evolution*, 71, 2661–76.

WATSON, N. L. & SIMMONS, L. W. 2010. Mate choice in the dung beetle *Onthophagus sagittarius*: are female horns ornaments? *Behavioral Ecology*, 21, 424–30.

WAYNFORTH, D. 2007. Mate choice copying in humans. *Human Nature—an Interdisciplinary Biosocial Perspective*, 18, 264–71.

WEDEKIND, C. 1994. Mate choice and maternal selection for specific parasite resistances before, during and after fertilization. *Philosophical Transactions of the Royal Society of London B—Biological Sciences*, 346, 303–11.

WEST, P. M. & PACKER, C. 2002. Sexual selection, tem-perature, and the lion's mane. *Science*, 297, 1339–43.

WITTE, K., KNIEL, N., & KURECK, I. M. 2015. Mate-choice copying: status quo and where to go. *Current Zoology*, 61, 1073–81.

ZUK, M. 1993. Feminism and the study of animal behav-ior. *Bioscience*, 43, 774–8.

CHAPTER 2

Female and Male Mate Choice: Similarities and Differences

2.1 Brief outline of the chapter

Females choose mating partners for three main reasons: direct benefits, indirect benefits, and compatibility, either genetic or social.

In this chapter I am not trying to look at all angles of mate choice, but to give a short overview of female choice to provide a basis for a comparison with male choice. This will highlight what studies are needed to reach a more complete picture of sexual selection. I would summarize the chapter like this: it's the ecology, stupid.

2.2 The role of sex roles

Biology has a number of major concepts and organizing principles. Sex is one of them. As a major concept it also has an important heuristic value: it directs the way we think about questions and influences how we ask questions and how we interpret findings. This heuristic value is of great importance, but sometimes it can lead to problems. Species, for example, is an extremely important concept in biology. Without an at least tacit agreement of what a species is, biology would have trouble articulating some of our major questions and interpreting some of our most important findings. Yet, within biology there are several competing species concepts, with different utility for different situations. The prevalent species concept—the biological species concept—has both history and baggage. It was conceived and formulated by eminent biologists including Ernst Mayr and Theodosius Dobzhansky in the 1920s and 1930s as part of the "modern synthesis" and defines a species as a group of organisms that can reproduce with each other, but are reproductively isolated from others (Mayr, 1940; Mayr, 1963; Milinski, 2006). This is a great definition and most importantly a great heuristic tool (De Queiroz, 2005), but it has limitations. For example, the biological species concept relies on having at least two sexes. What if an organism has only one sex (Schlupp, 2005)? Furthermore, the biological species concept is difficult to apply to microorganisms (Rosselló-Mora and Amann, 2001). The biological species concept also relies on mating behavior in its definition, but it is very hard to study reproductive behavior in species that are already extinct. Consequently, other species concepts have been proposed and are used in other fields (De Queiroz, 2005). There is an ongoing and lively debate around the species concept, which partly just reflects the complexity of living organisms, but it does illustrate that organizing principles can steer thinking and research. Among other things, the biological species concept emphasizes the cost of mating with members of other species and consequently deemphasizes the role of hybridization in evolution and ecology. It took the field some time to realize that hybridization between animal species is actually more common than we thought and can also be a driver of speciation (Grant and Grant, 1992; Mallet, 2007; Lamichhaney et al., 2018; Edelman et al., 2019). One might argue that organizing principles can also turn into assumptions we don't question anymore.

Another important organizing principle is sex. There is little doubt that the vast majority of metazoans have either large gametes or small gametes. Sometimes these are found in the same individual, but intermediate sizes of gametes essentially do not exist. Some fungi may be the only outliers. Here,

Male Choice, Female Competition, and Female Ornaments in Sexual Selection. Ingo Schlupp, Oxford University Press (2021). © Ingo Schlupp (2021).
DOI: 10.1093/oso/9780198818946.003.0002

several mating types, which could otherwise be viewed as sexes, can co-exist. The dynamics and evolutionary consequences of these mating types are not well understood but extremely fascinating. While in almost all metazoans gametes come in only two kinds, in fungi many different mating types can co-exist (Billiard et al., 2012), leading to thousands of "sexes" (Kothe, 1996). At the very least these fungi challenge the notion that sex is a binary or bimodal phenomenon (Billiard et al., 2011). Another example in this context might come from a species of bird. It has been argued that the white-throated sparrow (*Zonotrichia albicollis*), has effectively four sexes, namely two male kinds and two female kinds (Tuttle, 2003; Tuttle et al., 2016), but these may be more akin to mating types and still only present either large or small gametes. This is because of an inversion in one of their chromosomes that prohibits crossing over and effectively renders this chromosome a sex chromosome comparable with the mammalian Y chromosome.

Furthermore, although we associate sperm with maleness, sperm are not always small. The giant sperm found in some species of *Drosophila* are an example of this (Bjork and Pitnick, 2006; Lüpold et al., 2016).

A strictly binary view of sexes has important consequences at all levels of inquiry. This may be true not just for sexual selection but for all of evolutionary biology (Ahnesjö et al., 2020). We typically associate a syndrome of traits, including behaviors and a certain ecology, with small gametes and call them males, while we call individuals with large gametes females. The view that only two types of gametes exist is well founded in theory and supported by empirical evidence. However, the association with sex roles may be far less clear. In this sense, the roles these two sexes typically are thought to play may have led us to overlook some of the true complexity of female and male behavior and ecology. Females are typically choosy of their mating partners, but not always, and males are often not choosy, but sometimes they are. Have these patterns been overlooked and underappreciated? Maybe not overlooked, but probably underappreciated, as I will explore with a few examples later.

The ascribed sex roles of choosy females and indiscriminate males based on investment in

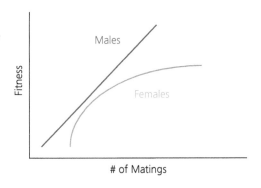

Figure 2.1   Bateman gradients differ for males and females (redrawn after Morrow, 2015). First appeared in MORROW, E. H. 2015. The evolution of sex differences in disease. *Biology of Sex Differences*, 6, 5. Open Access article material reprinted under Creative Commons Attribution 4.0 License.

gametes—essentially based on Bateman's work (Bateman, 1948)—have been corroborated many times. On the other hand, though, male mate choice has evolved as a strong force in some species where males make such a large investment in offspring that they become the limiting sex and females compete for mating opportunities with males. One has to keep in mind, though, that investment in offspring is only one path to the evolution of choosiness. Alternatively, for example, the adult sex ratio (ASR) has been suggested as driving this change in sex roles (Liker et al., 2013). Females may be aggressive among each other and evolve ornaments, traits usually associated with males. Hence, such species, including pipefishes and jacanas are often classified as sex-role reversed. The traditional sex roles are based on the idea of Bateman gradients (see Chapter 1) (Figure 2.1), where parental investment heavily influences which sex evolves to be the limiting, choosy one. Consequently, the traditional view of sex roles is historically fraught and, although it has been critiqued in the past (Kvarnemo and Ahnesjö, 1996), it persists as an essentially binary concept. It appears to reduce the existing variability to just four conditions: male acting as male, female acting as female, female acting as male, and male acting as female. But there is more than either only female choice or only male choice, as has been pointed out by several authors (Bondurianksy, 2009; Ah-King and Ahnesjö, 2013). There are many studies of mutual mate choice, which I will discuss later in Chapter 9. The view that female or male investment leads to only binary sex roles (which can flip, based on investment) is therefore problematic. Furthermore, using simple Bateman gradients has recently been discussed as overestimating the importance of traditional sex roles (Collet et al., 2014), but there is also strong evidence that sexual selection is indeed stronger in males (Janicke et al., 2016). Bateman gradients can also be similar for males and females in some species (Ursprung et al., 2011). Finally, a comparison of measures of sexual selection—and the Bateman gradient is just one of several—found that another index, the Jones index (Jones, 2009), is better than Bateman gradients at predicting the strength of sexual selection (Henshaw et al., 2016). Overall, Bateman's work and the influence it had are now perceived much more critically than in the 1980s (Hoquet, 2020). Yet the impact of this paper and the one by Trivers (1972) on how we view and construct sex roles is still heavy and the debate on the role and utility of Bateman gradients in interpreting sexual selection is still open (Arnold, 1994; Wade and Shuster, 2005).

In reality, in a large number of species both sexes invest more than just gametes in their offspring, and the playing field may be more level than previously thought. If investment into offspring selects for some degree of choosiness, we may be missing important aspects of sexual selection by only looking at the traditional sex roles of choosy females and unselective males. Species where both sexes invest, and both should be choosy to some degree, are the most interesting to me. In a seminal review paper, Bonduriansky used the term "partial sex role reversal" for such situations (Bonduriansky, 2001). But interestingly, species where only males choose and females compete for access to males are relatively rare (Petrie, 1983). I suggest we think of the situation where one sex only chooses and the other only competes as the extreme ends of a continuous distribution, with some degree of choice and competition in both sexes in between.

To evaluate the importance of sex-role reversal, I will take a look at the best-understood examples of sex-role reversal. These are probably the pipefishes, a taxonomic group within the Syngnathids, which are a family of marine fishes also including seahorses. They exemplify how important the ecology

of parental investment is for the evolution of choosiness. They are typically viewed as supporting evidence for the evolution of sex roles because in pipefishes the sex roles are reversed. But they also seem to cement the view of fixed, binary sex roles (Vincent et al., 1992). In Syngnathids males have evolved paternal care, and many have a unique brood pouch that is used to brood the offspring (Wilson et al., 2001). While all species show paternal care, many are monogamous, with little female competition going on. In some species, however, males are polygamous, and females compete for males (Rosenqvist and Berglund, 2011). They act essentially as males would act in "traditional" species (traditional used here as defined by Kvarnemo and Ahnesjö, 1996). They are larger than males and size is important in male choice and female reproductive success (Braga Gonçalves et al., 2015), they court and aggressively compete for males (Berglund et al., 1993), and show ornamentation (Berglund et al., 1997; Berglund and Rosenqvist, 2001). Males select females carefully, based—among other things—on parasite load (Partridge et al., 2009), which can also induce postcopulatory male choice. Because all of this is parallel to female behaviors in many "traditional" species, it is typically viewed as powerful evidence for the generality of sexual selection. It clearly shows that similar selection pressures can lead to comparable adaptations. One species, *Syngnathus typhle*, the broadnosed pipefish, is especially well studied (Berglund and Rosenqvist, 2003), but the group as a whole is an important model system in evolutionary biology (Mobley et al., 2011; Braga Gonçalves et al., 2015). In one study of *S. typhle*, Jones and colleagues (2000) measured the Bateman gradient, and provided evidence that sexual selection is indeed operating strongly on females, but—somewhat surprisingly—also weakly on males (Jones et al., 2000). This supports the notion that both sexes can be under sexual selection at the same time.

A central thesis of this book is that mate choice can evolve also in the sex that invests less in any mating system, largely based on differences in female quality, male investment, and sex ratio (Edward and Chapman, 2011). Likely, the ecology of investment can drive the evolution and maintenance of mate choice. Applied to pipefishes this predicts that not only males, but also females may be choosy—if the conditions are right. This was actually found to be the case (Berglund et al., 1986) in two species of pipefish, *S. typhle* and *Nerophis ophidion*. In these species both males and females preferred larger partners. This was further supported—albeit a little indirectly—by Sandvik and colleagues. In their study they found that both males and females pick partners that affect offspring quality positively (Sandvik et al., 2000). Interestingly, size was not a significant factor in female choice in this particular study.

In amphibians the green poison frog, *Dendrobates auratus*, was suggested as sex-role reversed (Trivers, 1972), but this was not confirmed (Summers, 1989), although female–female competition and very active female behavior during courtship was found (Summers and Tumulty, 2014).

Other important species that are considered to be sex-role reversed are found among Nordic shorebirds (phalaropes) and tropical jacanas (Figure 2.2). The two European species of phalaropes (*Phalaropus fulicarius* and *P. lobatus*) breed in the High Arctic and live a largely pelagic life on the open ocean and are thus somewhat difficult to study. However, the Nearctic Wilson's phalarope (*Phalaropus tricolor*) is widespread in the prairies of Canada and the northern United States. Phalaropes and jacanas are in the same suborder (Scolopaci) of birds in the order Charadriiformes, but sex-role reversal seems to have evolved at least twice independently (Gibson and Baker, 2012), potentially driven by the ASR (Liker

Figure 2.2 Male jacana (*Jacana jacana*) at the nest (photo credit: Sarah Lipshutz).

et al., 2013), an argument we will revisit in Chapter 6. Consequently, they show male mate choice (Whitfield, 1990; Schamel et al., 2004) and females compete for males (Colwell and Oring, 1988). Females are somewhat more ornamented than males, but both sexes have a much more cryptic appearance when they are not breeding (Höhn, 1965). I will come back to female ornamentation in Chapter 8. In jacanas (*Jacana jacana* and *Jacana spinoza*)—and similarly to phalaropes—males take care of the chicks, and the larger females compete for males (Butchart, 2000; Emlen and Wrege, 2004) (Figure 2.2). Female competition can lead to takeovers of males and infanticide by the new female (Emlen et al., 1989). The family Jacanidae has eight members, all of which are sex-role reversed, with one exception, the lesser jacana, *Microparra capensis*, which is not (Hustler and Dean, 2002). It seems that within the Jacanidae sex-role reversal is ancestral and the traditional sex roles are a derived character state (Whittingham et al., 2000).

Eens and Pinxten (2000) reviewed cases within vertebrates for sex-role reversal and considered some fishes, amphibians, and birds to match the description. Interestingly, reptiles and mammals are absent from the list (Eens and Pinxten, 2000), although it is not clear why. More importantly they asked if there are hormonal mechanisms that unite the known examples of sex-role reversal, such as high testosterone in females, but did not find a conclusive pattern.

In addition to vertebrates, there are several other examples of apparent sex-role reversal in invertebrates. One example is a group of small cave-living insects in the genus *Neotrogla* (Yoshizawa et al., 2018b), where females are competing for males, which provide sperm for nutritional purposes. In addition to showing behavioral reversal of sex roles, this insect has also evolved a penis-like copulatory organ in females (Yoshizawa et al., 2018a, 2019) (Figure 2.3). This may play a role in coercing males to deliver sperm. In this case, too, ecology is crucial to understanding the mating system: the dry caves in Brazil that these insects inhabit are low in food and females seem to require nutritional sperm from multiple males to reproduce.

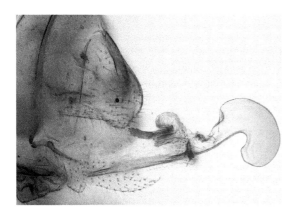

Figure 2.3 Female *Neotrogla curvata* penis (photo credit: Kazunori Yoshizawa).

Figure 2.4 Two spotted goby (*Gobiusculus flavescens*) female courting a male, a behavior called "hook and glow" as the female bends her body and contracts melanophores on her back (photo credit: Andreas Svensson).

In the bigger picture, male investment in offspring and an operational sex ratio (the ratio of sexually receptive males to females; OSR) that makes males the rarer sex seems to set the stage for males showing fully developed, exclusive male choice. Yet, overall few species seem to fully fit the bill of complete sex-role reversal. And the taxonomic distribution is difficult to interpret. What is similar between Nordic phalaropes and tropical jacanas that let those species evolve full male mate choice? One would think that male parental care and the evolution of brood pouches in pipefishes set the stage for the evolution of male mate choice, but pipefishes are not the only fishes showing this

kind of brood care. Actually, in fishes many taxa, such as cichlids, show extensive male parental care, but not sex-role reversal. Perhaps we are not asking the right questions.

Work on another group of small marine fishes, gobies, illustrates further just how important the role of ecology is for the evolution of male mate choice and also demonstrates how dynamic male and female choice can be (Kvarnemo and Ahnesjö, 1996; Forsgren et al., 2004; Amundsen, 2018; Heubel, 2018). In this case, the sex roles that often seem so fixed turn out to be highly flexible. In the two-spotted goby *Gobiusculus flavescens*, a small marine fish found along the west coast of Sweden (Figure 2.4)—among other gobies—both females and males can be choosy, based on the ecological and social environment. The key variable that seems to drive these decisions is the OSR. This concept goes back to the 1970s (Emlen and Oring, 1977) and has been important in understanding mating systems. It was criticized as lumping too many individual variables into just one (Kokko and Johnstone, 2002), especially parental investment. Overall, however, the concept is widely used in the field. For male mate choice it means that whichever sex is rare at a given time should be choosier than the other one. As OSR can vary in space and time, so can choosiness. In some gobies, for example, the females start out choosy early in the season when more males are present than females, but toward the end of the breeding season, females have to compete for the increasingly rare males (Amundsen, 2018). In the late season, consequently, males become choosy. All of this happens within a few weeks, with fairly dramatic consequences for the mating system. A related, but separate, concept is exploring the role of the ASR (proportion of adult males in the adult population) in the evolution of mating systems and this also concludes that male-biased ASR favors the evolution of sex-role reversal (Liker et al., 2013) (Chapter 6). In the context of the studies on gobies, let me play devil's advocate for once: what if—for whatever reason—scientists only studied the goby mating system early in the breeding season? We would have missed a very important aspect of sexual selection in gobies and might firmly conclude that these gobies have a mating system based on female choice only.

2.3 Male mate choice based on female quality

Females and males can differ significantly in quality, and these differences are important in mate choice. In the case of male mate choice this often manifests itself as differences in female fecundity. At least it seems as if a fecundity benefit is the first explanation that comes to mind when we detect a form of male mate choice, but is this really always the best answer or just the most widely accepted one? Just like males, females can differ in other ways, too. We know from female choice that many factors influence it, ranging from the abiotic and social environment to the perceived quality of the mate, to the internal state of the chooser. And just like female choice, male choice begins with determining whether a potential mate is from a species with which it is possible to pass on their genes. Mating is mostly local and, in reality, the pool of available mates is limited in time and space. Humans, for example, seem to respond to such limitations in a number of ways: individual men can mate with less preferred partners, or they can increase the search radius—thus incurring higher search costs—or they can wait for a more suitable partner (Jonason et al., 2019). All these strategies have their own costs, of course, and they are gender specific.

2.4 From species recognition to mate choice

Sometimes female choice is presented as a cascade of decisions, starting with identifying the right species, and eventually exercising a preference for a particular mate. Often there is no conflict between these domains of mate choice, and if there is a conflict, it can be resolved if females evaluate between-species traits using different criteria as compared with within-species traits (Rosenthal and Ryan, 2011). Indeed, some early work on male mate choice was conducted in the context of species recognition. Some examples of this were reviewed by Schlupp (2018). Modern work is tackling this problem by looking at both female and male choice (Mendelson and Shaw, 2012; Roberts and Mendelson, 2017; Mendelson et al., 2018). It seems plausible that

males are selected not to waste their effort with females of a different species as mating with them may result in no offspring. One could argue that often the cost for males of mating with the wrong species is relatively small. Yet, sometimes—if we stick with this analogy—the return is zero, and even if the cost is small, there should be selection favoring individuals that avoid matings which provide no fitness return. This is one example of how important species recognition in mating behavior is. However, perfect choice is unlikely to evolve because it would be too costly (Kokko et al., 2008; Heubel et al., 2009). This is, for example, the case in a number of species that reproduce via sperm-dependent parthenogenesis (Dawley, 1989). They are essentially sexual parasites (Hubbs, 1964). In these species females form eggs that require some kind of fertilization, but the male genes are typically not incorporated into the germ line (Suomalainen et al., 1987). The females depend on obtaining sperm, but the males typically gain nothing from such matings, except indirectly via mate copying (Schlupp et al., 1994; Heubel et al., 2008). Hence males should evolve species recognition and avoid mating with such females (Gabor and Ryan, 2001; Gabor et al., 2005).

From the female perspective such a mating system is combining the worst of both worlds: they face the costs of asexuality like the lack of recombination and at the same time many of the costs of sexuality because they need to mate with males. Both pure parthenogenesis (as found in many invertebrates and some vertebrates) and pure sexual reproduction (found in the vast majority of metazoans) are more common than sperm-dependent parthenogenesis. The latter seems to be fraught with both the costs of asexuality, because they have no recombination, and the costs of sexuality, because they have to mate to obtain sperm. A mating system like this is found in Amazon mollies (*Poecilia formosa*), a gynogenetic fish (Hubbs and Hubbs, 1932). In gynogenesis male sperm is necessary to trigger embryogenesis, but the male DNA is not incorporated into the embryo. This fish is also of hybrid origin and males of the two sexual and ancestral species provide sperm for the Amazon mollies (Schlupp, 2005). To complicate things further, Amazon mollies are livebearing fishes and have inner fertilization, which requires sperm transfer from males using a modified fin, the gonopodium. For the males involved in this, there is a choice between females of their own, sexual species and the Amazon mollies. Clearly, they should prefer the conspecifics, and they often do. There are a number of complicating factors, though. First, theory shows that perfect male choice is costly to evolve (Kokko et al., 2008). Maybe surprisingly, it is evolutionarily stable to allow for some mistakes in male mate choice. Second, even though matings with Amazon mollies are a poor choice for the males, they are essential for the Amazon mollies and they have evolved to obtain the needed sperm.

Evolutionarily speaking, this is reminiscent of the "life-dinner principle" introduced by Dawkins and Krebs (Dawkins and Krebs, 1979) to highlight the different selection pressures on prey and predators: if a predator misses a meal, it may not be the end of it, but if the prey *is* the meal, it ends all future reproduction for the prey. Maybe because of this, Amazon mollies will, for example, interrupt an ongoing mating of a male with a sexual female and try to be inseminated instead (Schlupp et al., 1992; Marler et al., 1997). Nonetheless, there is very solid evidence for male mate choice in this mating system (Schlupp, 2009, 2018). Interestingly, this is a mating system in which social influences on mate choice, like mate copying (Schlupp et al., 1994; Schlupp and Ryan, 1997; Witte and Ryan, 2002) (see Chapter 1) and audience effects, are very strong (Plath et al., 2008). Similar mating systems exist in a number of other vertebrates and invertebrates (Suomalainen et al., 1987). The example of the Amazon molly highlights that many conditions exist that predict the evolution of male mate choice, although it may not always be easy to detect it. Potential mating partners can differ in many different aspects, not just the binary correct or wrong species. I will return to many of these aspects in separate chapters after providing a taxonomic overview of where male mate choice has been reported in Chapter 3.

Often species recognition and mate choice are viewed as a continuum (Ryan and Rand, 1993). One of the arguments for an important role of species recognition is that hybridization can be very costly to females. Recent work on hybridization, however, has shown that hybridization between animal species is far more common than previously thought, and has led to massive gene flow between species (Mallet, 2007; Cui et al., 2013). If hybridization is less costly than we think, then maybe selection against heterospecific matings is less strong.

2.5 Female and male choice

Male mate choice is not simply the inverse of female choice, but there are similarities. The main difference is that male choice does not have the same potential evolutionary outcomes as female choice. Male choice is much less likely to select for traits that negatively impact female survivorship, which is one of the hallmark features of female choice. Many of the exaggerated traits we consider ornamental in males are detrimental to male survival but have evolved nonetheless, because they are advantageous in sexual selection. It seems less likely that females would evolve equally risky strategies and show equally costly traits simply because of their higher investment in gametes. Consequently, it seems as if male choice has different effects on females than female choice has on males (Fitzpatrick and Servedio, 2017, 2018). This argument is both intuitive and supported by theory. The same reasoning holds true for the evolution of female ornaments. While male ornaments often evolved to elicit positive responses from females, and at the same time have costs for males, a similar evolutionary outcome for females seems unlikely. Because females are bearing the costlier gametes, any trait that would negatively affect their survival is much less likely to evolve. Despite this fundamental difference, I argue in a later chapter that male choice is not trivial, and female ornaments (and armaments) are more than pleiotropic effects of male ornaments.

2.6 Limitations of choice

Before addressing female and male choice in a little more detail, I want to make clear that preference and choice are not all that determines reproductive outcomes. Foremost, sexual conflict is interfering with choice. Members of the limiting sex are often faced with adaptations in the limited sex used to undermine choice. The partners involved in mating may have vastly different optimal outcomes, generating sexual conflict (Chapman et al., 2003; Parker, 2006). Such conflict can be over the number of matings, duration of mating, or other aspects. After the choosing and mating phase, conflict can be over subsequent investment in offspring. Males can undermine female choice by forcing females to mate with them. All of these manifestations of sexual conflict interfere with preferences and choice. How big the mismatch between preference and actual outcome is, is difficult to determine, but we know there must be some mismatch. This imperfection might be viewed as a reflection of the many costs of choice and accepting a less than optimal mating partner as the resulting compromise. The costs of female choice are well documented. Often female choice is diminished when the female is facing a substantial risk, such as predation. There are fewer examples of males adjusting their choice behavior relative to a cost. One example is males of the Pacific blue-eye (*Pseudomugil signifer*), a small fish from the east coast of Australia, which show a preference for larger females, but modify that preference when they are forced to swim in a current. Under those circumstances they are more likely to associate with a smaller, less fecund female (Wong and Jennions, 2003).

Ideally, choice would be based on perfect information gathered by the chooser, but this is never the case. For example, choice is heavily influenced by temporal factors, such as limited time to make a decision. Mating may be the most important thing a chooser ever does, but it is not the only aspect of life that requires their attention at any given time. Consequently, mating decisions are influenced by some inner state, such as hunger (often regulated by hormones), health, or the presence of competitors and predators. The social environment in general is extremely important, sometimes locking in early experiences (Quintana et al., 2018). Also, mating preferences change over time and may have innate and plastic components (Plath et al., 2019). Furthermore, mate preferences in general often have a large learned component. This has been well documented in song learning and song preferences in birds (Beecher and Brenowitz, 2005; Brenowitz and Beecher, 2005) with recent studies highlighting the important role of the social environment (Lachlan et al., 2018; Searcy and Nowicki, 2019).

Not only the conditions and the quality of the choices vary with time and space, but also the inner state of the chooser, as also demonstrated in detailed studies of the hormonal mechanisms of song learning (Boogert et al., 2018). All of this leads to an ever-changing cost–benefit matrix for mating decisions. Consequently, we expect results from mate choice experiments to reflect this complexity.

Furthermore, the time of choice—for example relative to the mating season, if there is one—is important. Early in the season, a chooser may be more likely to reject poor-quality mates. This can be influenced by several factors. The probability of a better mate arriving may still be high, the OSR may still be in favor of the chooser, and the internal state of the chooser may not require mating yet. A related, newer concept, looking at the ASR (Ancona et al., 2017; Carmona-Isunza et al., 2017), makes predictions based on the ratio of all adults present in a population (e.g., Grant and Grant, 2019). Using the ASR shifts the focus a bit from just looking at the breeders and sexual selection *sensu stricto* to including all adults, with a more inclusive view as also proposed in social selection (Lyon and Montgomerie, 2012).

In addition to variation over time, there is also important variation in the spatial distribution of potential mating partners, which can lead to less than perfect choice. Furthermore, mate choice is typically local, with mate assessment only possible on that scale. For treefrogs (*Hyla chrysoscelis*) this is a radius of about 2 m, for other animals this may be much smaller or much larger (Morris, 1989). In humans, however, increased mobility and the use of dating apps may be reshaping what is local (Quiroz, 2013). Many other factors, such as for example hunger, can influence mate choice: hungry females show reduced choosiness and hungry males show reduced courtship vigor (Billings et al., 2018). Furthermore, processing the information needed to make mating decisions is likely to be influenced by distractions and cognitive load issues (Sweller, 2011). Assuming that processing power is limited, mating decisions may be compromised by attention paid to other aspects of an individual's social environment. One potential example is distraction by an audience, which was hypothesized to negatively influence mate choice in fishes (Plath and Schlupp, 2008). This is also true with regard to the abiotic environment. One has to highlight the complexity of the situation and the interconnectedness of factors. The red shiner (*Cyprinella lutrensis*), for example, a common fish inhabiting lowland rivers in the American Midwest, responds to increased turbidity with a red shift in the expression of a visual pigment (Chang and Yan, 2019), which is in turn likely to influence the mating behavior and social environment in this fish with strong red ornaments. Finally, one has to acknowledge that individuals may simply make mistakes for various reasons and mate with a less than perfect partner.

Taken together, this means that the ideal mate may simply not be available at the place and time when choice happens, and animals have evolved decision rules to adapt to this. But, within these confines, preferences and choice are immensely important for the reproduction of individuals.

2.7 Female choice

Mate choice is incredibly complex and complicated (Rosenthal, 2017). Yet, there are still a number of fundamental questions to be resolved, such as whether mate choice is mainly rational and transitive (Hemingway et al., 2019; Ryan et al., 2019), and what role the ecological and social environment is playing (Danchin et al., 2004; Ryan et al., 2007). Consequently, it would be impossible to give more than a somewhat partial overview in this chapter. Instead, what I would like to do here is highlight some of the core aspects of female choice with the purpose of trying to find mirroring aspects in male choice. Females choose particular males because of at least one of three adaptive advantages. First, they may provide direct benefits to the female, second there may be indirect genetic benefits, and third, they may be most compatible with the females. In addition, males may evolve traits that trigger previously untapped preferences and thus deliver none of these suggested advantages. This is basically Darwin's argument from sexual aesthetics.

Direct benefits can come in many forms, for example, as food items provided during courtship, or access to other resources for the females such as territories, nesting sites, shelter, and so on. Anything

that leads to more or better-equipped offspring would be in this category. Evidence for this is abundant and very convincing. Indirect benefits are more difficult to document, and evidence for them is less clear than evidence for direct benefits. Indirect genetic benefits can be present when females gain a benefit for their offspring other than a direct benefit. The argument is intuitive, and essentially says that attractive males will sire attractive sons that have a better chance at reproducing in the next generation. This is often called the "sexy son hypothesis" (Weatherhead and Robertson, 1979).

The third adaptive advantage, compatibility, is conceptually a little different. This idea posits that females benefit from picking the best matching male, but not necessarily one that is preferred by all other females. The latter is an assumption for both direct and indirect benefits. One mechanism for achieving compatibility is mating with a male that has a matching genotype that leads to maximum major histocompatibility complex (MHC) diversity in the offspring. On closer inspection, preferences for MHC are actually quite complex (Milinski, 2006), but the important difference to the previous two mechanisms is that females do not have to show similar preferences within a population. Compatible mates, but this time socially compatible, are also important in species where pairs form long-term relationships, such as in humans (Buss and Schmitt, 2019).

Finally, females may choose certain males because they elicit a strong sensory response in them. A pre-existing bias can attract females to males that somehow match this bias (Ryan and Cummings, 2013). This may not be for adaptive reasons, which distinguishes this mechanism from all the other ones. Of course, all of these mechanisms are not mutually exclusive.

Below I will discuss these concepts in a little more detail, to set up a comparison with male mate choice.

2.8 Female choice for direct benefits

Examples of direct benefits to females provided by males are abundant, and the effects appear to be quite clear. But there is some debate about the magnitude of effects relative to that of indirect benefits (Møller and Jennions, 2001). I want to discuss a few examples here, not only because of their importance in the context of female choice, but also because they will be relevant later, when I discuss male choice. Direct benefits provide a material advantage for the female, increase the number or quality of the offspring, or increase their chance of survival. Clearly, quality and survival chances are more difficult to measure, but they may still be relevant. Many species, for example, show some kind of nuptial feeding, in which a nutrient, food, or water is provided to the female by the male. In several insects for example, the spermatophore is consumed by the female and becomes the nuptial gift (Lehmann and Lehmann, 2016). Others bring gifts for the female (Engqvist, 2007; Sakaluk et al., 2019), and in some cases parts of the male (Alexander and Otte, 1967) or the male itself are the nuptial gift (Fromhage et al., 2003).

In modern humans, men may invest heavily in offspring and partner by providing various resources. However, not all men meet this societal expectation. In many societies this investment is nowadays measured in money, which somehow reduces the phenomenon to just one currency, while clearly nonmonetary contributions, such as time and affection are important, too. This leads to a situation, somewhat unusual in mammals, where both genders, women and men, make investments. Yet on an individual basis, there are major differences in how and how much men invest in their offspring. But overall, these investments are substantial. Although difficult to quantify, I think that on balance, women invest more in any given offspring. In many societies, however, male investment is not only part of the ethical code, but also part of the legal system. Men, and more generally parents, are obliged to provide for their offspring until they are deemed independent. This, of course, depends on the societal context and can lead to discrepancies between the legal age of independency and actual independence. In many Western societies the age of legal adulthood begins at 18, whereas often financial independence comes at a later age. Because female choice often happens before the female can know with certainty that her partner is a good provider,

characteristics indirectly associated with this are preferred by women (Buss, 1989).

Male investment in the mating partner and/or the offspring is quite common. As discussed in Chapter 1, there is a continuum in male investment that reaches from no male investment beyond the minimum consisting of sperm and the fluid that is used to deliver it, as found in many fishes and amphibians, to male investment that is so heavy that it tilts the scale and makes males the rare sex, such as in pipefishes and jacanas (Figure 2.2). In some groups, like cichlid fishes, several of these strategies are found in closely related species. In cichlids, a speciose family of fishes, some species show no male investment beyond providing semen, while others provide intensive biparental care or paternal care. For example, this can come in the form of mouthbrooding, where the male forsakes feeding, and holds first the eggs and then the larvae in his mouth cavity (Balshine-Earn and Earn, 1998; Goodwin et al., 1998). This is a very efficient mechanism for providing protection for the offspring.

For practical purposes, I distinguish between investments made in the offspring via the female and direct investments in the offspring. Females can receive nuptial gifts as they are incubating. These gifts can be food, or water, helping the female to survive and produce eggs. Such provisioning is found in many insects, including some katydids (Gwynne, 1986; Simmons, 1995), but may not be universal to provide direct benefits for females. In a study using dance flies (*Rhamphomyia sulcate*), LeBas and colleagues found that small males with small gifts are favored in sexual selection (LeBas et al., 2004), potentially owing to better maneuverability of males. In sexually cannibalistic species males can benefit from becoming a food item themselves. This is common in some spiders and praying mantises (Maxwell et al., 2010). In these cases, the male probability of finding a second mate is so small that they provide their own body as food for the female. These mating systems are also characterized by strong size dimorphism, with males typically being much smaller than females. In several species the size of the gift determines the amount of time for which a male is copulating with a female, and correlates with the amount of semen transferred, which in turn correlates with paternity. Matching genitalia can have a similar effect (Holwell et al., 2010). All of these are examples of males providing a material benefit to the female and indirectly to the offspring.

Actually, nuptial gifts can also be used to highlight another important aspect of mate choice, namely sexual conflict. What if males used nuptial gifts or semen to manipulate female reproduction? This would reflect the sexual conflict briefly discussed before. As males and females have different optima, for example, for the number of matings, this generates sexual conflict. Therefore, the gifts from males may not be a kind contribution by the male to lessen the burden for the female, but an attempted investment in their own offspring. If there is an evolutionary path to manipulating females such a behavior may arise (Crean et al., 2016). Similarly, the ejaculate can provide important substances for the female—in addition to sperm— but may also manipulate females. In fruitflies (*Drosophila melanogaster*), male ejaculates can lower the fitness of females (Chapman et al., 1995). For the males this is, however, adaptive because other males are unlikely to mate with that particular female. They secure their paternity at the cost of the female. Of course, if females can evolve to manipulate male investment that may also evolve (Holland and Rice, 1999).

In addition to providing nutrition for their mates, males can also provide directly for their offspring. Maybe one of the most obvious examples is feeding the offspring. But even this seemingly simple scenario is more complicated on a second look. Offspring feeding is widespread among birds, and both males and females can participate in this activity. From the standpoint of compensatory male investment, offspring feeding can be important, but if females also feed the offspring, the existing gap between male and female investment will remain open. In mammals, such additional female investment, lactation, is the hallmark of the group. In many species, females can provide unfertilized eggs, which are consumed by their offspring. Such trophic eggs are widespread, for example, in insects (Baba et al., 2011) and some frogs (Dugas et al., 2016). Frogs are especially interesting in this context because they have also evolved a form of male parental care: males of some dart poison frogs carry tadpoles on their back, sometimes even providing trophic tadpoles to their cannibalistic offspring (Schulte and Lotters, 2013). This represents an interesting mix of female and male investment.

In birds feeding of chicks by males is very widespread. Often starting with shared brooding of the eggs, males of many species provide food for their offspring. Because it is relatively easy to quantify, this form of paternal care is well documented. One popular hypothesis is that males scale their effort according to their likelihood of paternity (Møller and Birkhead, 1993; Møller and Cuervo, 2000), but this view has not been supported by some analyses (Schwagmeyer et al., 1999; Sheldon, 2002).

Another mechanism how males can provide direct benefits and care is to protect the offspring from the environment, including predators. This type of behavior is very widespread and can range from mouthbrooding in fishes, to providing and defending shelter in the form of cavities, to carrying the offspring on their back (Woodroffe and Vincent, 1994; Tallamy, 2000).

Together, it is clear that direct paternal investment is very common and has evolved many times.

2.9 Indirect benefits

Females may also choose in the absence of direct benefits, when males only provide fertilization for the females and nothing beyond. As discussed above, in some species with internal fertilization the ejaculate may provide some direct benefit if the female can somehow resorb it, but that benefit is likely to be quite small. By contrast, there are potential costs to receiving ejaculates, for example infection with sexually transmitted diseases (STDs) (Knell and Webberley, 2004), or male attempts at manipulation (Chapman et al., 1995). Indirect benefits, on the other hand, come in the form of a genetic contribution to the offspring and may be the adaptive advantage for exercising choice (Greenfield et al., 2014).

Plausible support for "good genes" was for example reported in gray treefrogs (*Hyla versicolor*) (Doty and Welch, 2001), but one of the best examples was presented by Petrie and her colleagues. In her studies on peacocks and peahens she found that females indeed gain from mating with highly ornamented, healthy males via higher offspring survival and growth (Petrie et al., 1991; Petrie, 1994). While these results were independently confirmed (Loyau et al., 2008), one study was critical of the findings and unable to repeat the results (Takahashi et al., 2008), leading to a debate over the generality of the original findings (Loyau et al., 2008). Overall, it seems fair to conclude that indirect benefits may play some role, but it seems rather limited (Achorn and Rosenthal, 2020).

2.10 Compatibility

I started this section by pointing out that selection to mate with the right species is based on genetic compatibility. Now I am going to look at genetic (and social) compatibility within a species. The indirect benefit and direct benefit concepts are unified by assuming that there is a "best" partner that choosers can choose. This makes studying average preferences for populations very useful and allows us to predict how selection would act on a population. However, sometimes what is best for one chooser may not even be good for another chooser. Sometimes females show preferences for the best match. The best developed example is widespread preferences for compatible genes of the MHC (Milinski, 2006, 2014). This set of genes plays an important role in immune defense of vertebrates and there seems to be an optimal number of loci to have. Females in some species are able to smell the MHC status of a potential mate (Jaworska et al., 2017) and show preferences for mates with whom their offspring will have the best possible MHC diversity. This has been documented for example in fishes (but see Promerova et al., 2017), birds (Baratti et al., 2012), badgers (Sin et al., 2015), horses (Burger et al., 2017), mandrills (Setchell et al., 2016), and humans (Qiao et al., 2018; Wu et al., 2018), to name some of the taxa.

The first study on MHC-based choice in humans came from a Swiss research group (Wedekind et al., 1995) using t-shirts worn by men to determine female preferences based on smell. The result supported the idea that women unconsciously used smell to determine compatible men who would produce offspring with high MHC variability. This work on selection for matching genotypes

intersects with work done in another field, namely on cryptic choice. In humans there is evidence for an interaction of sperm and cervical mucus and combinations that differ the most in human leucocyte antigen (HLA) show the highest sperm viability. This suggests that female choice for dissimilar partners also happens after copulation (Jokiniemi et al., 2020). Some studies following up on the original t-shirt study have provided support for this MHC-based female choice, but a meta-analysis comparing multiple studies at the same time found only weak support for MHC-based choice (Kamiya et al., 2014). Another study points toward limited statistical power (Hoover and Nevitt, 2016) in the original studies, a problem that might be more common than assumed. Interestingly, men can tell the difference between women of different MHC type, but this information is not very important in mate choice (Probst et al., 2017). Overall, however, support for an important role of MHC in human mate choice may be somewhat limited (Havlicek and Roberts, 2009).

Another form of choice for compatibility is selecting against costly or detrimental genotypes. One example is the *t* haplotype in the house mouse (*Mus musculus*), which—if homozygous—leads to in utero mortality. Females prefer not to mate with such males, thus avoiding a fecundity cost (Manser et al., 2015). Male choice in such a context is unknown. Furthermore, in a hybridogenetic fish from the Iberian peninsula, *Squalius alburnoides*, females seem to prefer males that are hybrids, and choice is partly based on the genome of the choosing female (Morgado-Santos et al., 2018).

Mate choice based on compatibility is interesting for a number of reasons, but to me the most intriguing part is that choosers are not necessarily predicted to all prefer the same individuals. While this is an underlying assumption for direct and indirect benefits, it is different for compatibility. Under direct and indirect benefit scenarios, we would expect some measurable trait to emerge that all preferences converge on. Compatibility does not exclude this, but also does not necessarily predict this pattern. A male that is a good match for female A may be less good for female B. This has important consequences for the interpretation of data on

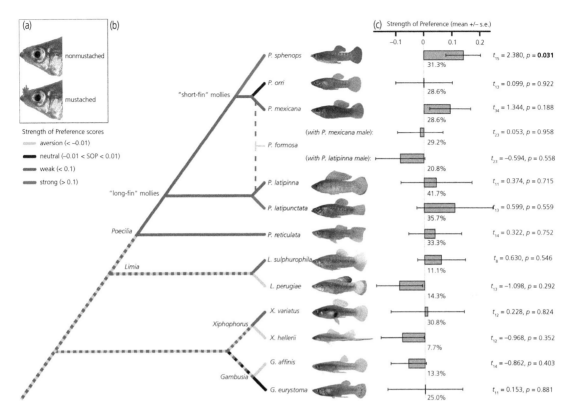

Figure 2.5 Individual variation in preferences for a novel trait in several species of livebearing fishes. Reprinted by permission from Springer Nature. *Behavioral Ecology and Sociobiology*, 65, 1437–45. 'Mustached males in a tropical poeciliid fish: emerging female preference selects for a novel male trait' by MCCOY, E., SYSKA, N., PLATH, M., SCHLUPP, I., & RIESCH, R. © 2011. Original artwork from Rudy Riesch.

preferences. If a whole population is expected to prefer the same trait values, individual variation is mainly noise. If females have individualized preferences, population means for trait values and preferences become more difficult to interpret. In an extreme case, individual preferences for traits may lead to binomial distributions of preferences and cancel each other out on the population level. As a side note, I want to mention that even in scenarios where we assume uniform preferences, there can be considerable variation in female preferences. In a study evaluating multiple species for their response toward a novel stimulus, McCoy and colleagues found that in all species of livebearing fishes studied, some individual females had a preference for the novel trait (a mustache; Schlupp et al., 2010), even when the preference was not statistically significant on the population level (McCoy et al., 2011) (Figure 2.5).

2.11 Assortative mating

Somewhat related and also interesting is assortative mating, where choosers may choose partners that share an important characteristic, such as size or color. Assortative mating might be a consequence of seeking socially compatible mating partners. While generally thought to be evolutionary important, a recent meta-analysis in amphibians found relatively little importance of assortative mating (Green, 2019). Another meta-analysis also revealed that assortative mating may be less important, and that positive publication bias may contribute to the impression that assortative mating is important (Jiang et al., 2013; Wang et al., 2019).

An interesting mechanism for assortative mating is suggested by some theoretical work. Costs and benefits of mate choice are different for different individuals. This leads to an important question:

should less competitive males prefer to mate with lower-quality females (Venner et al., 2010)? If so, this might be a mechanism for assortative mating. It also raises another important question. If, for some males, mate choice is too costly relative to the benefit, but not for others, we would predict populations of males in which some males are choosy, but others are not. Given the current methods we use to study mate choice, it is quite possible we would miss documenting male choice in such populations.

Sometimes females prefer males from the same region, which can contribute to population isolation, which in turn can lead to speciation. Furthermore, if, for example, members of a population have divergent preferences for, for example, food plants, this can even be the starting point for ecological speciation (Nosil, 2012). This can work via direct preferences for a mating partner, as shown in cichlid fishes from Lake Victoria (Carleton et al., 2005; Maan et al., 2006), or via a preference for a shared resource, such as a host plant in stick insects (Nosil et al., 2002). This relationship has been discussed in detail by Nosil and colleagues (Rundle and Nosil, 2005; Nosil, 2012).

2.12 Sensory bias

Finally, females may also prefer males for no adaptive reason at all, or because a preference is adaptive in another domain, such as exploiting sensory and cognitive systems outside of mate choice. Animals may be prone because of pre-existing sensory biases to prefer some males over others, without deriving any benefit from this (Ryan and Keddy-Hector, 1992; Basolo and Endler, 1995; Ryan and Cummings, 2013). It is plausible that signalers may exploit preferences in the receivers that evolved in a different context, such as feeding (Rodd et al., 2002), or for unknown reasons (Plath et al., 2007; Makowicz and Schlupp, 2013). I believe that a potential role for sensory bias in male mate choice has not been explored.

2.13 Proximate mechanisms and sensory biology

Selecting a mate involves much communication. Both the proximate and the ultimate side of this have been reviewed thoroughly (McGregor et al., 1999; Maynard Smith and Harper, 2003; Searcy and Nowicki, 2005; Bradbury and Vehrencamp, 2011). Clearly, female choice involves much more than just the ultimate, evolutionary outcome. There is a whole, very important proximate side to it. In order to pick a mate, choosers must have mechanisms in place to sense and evaluate them just as the signalers must evolve ornaments that provide information to the chooser. The choosers must be able to perceive the information provided by males, and process that information and come to a conclusion concerning the quality of the mate. There is no *a priori* reason to assume that males and females will have much different sensory systems, although this seems possible and was suggested for tungara frogs, *Physalemus pustolosus* (Hoke et al., 2010). Within any given species there may be differences in how sensory systems are used. And there may be changes over time, too.

One important outcome of studies of sensory biology is that sensory worlds can be fundamentally different between species. Even between and within individuals, sensory input can differ a great deal. Consider aging in humans: over time we lose the ability to hear certain frequencies, starting with higher frequencies and clearly correlated with age. A signal that is good for a young person may be inaccessible for an older person.

More importantly, between species there are major differences in how they perceive their environment. This allows the existence of private channels, in which animals can communicate within a species, while shutting out other species. Such private communication can also occur within species, but mostly sexual signals evolved to be easily recognized.

2.14 Cost of mate choice

The benefits of mate choice have been discussed above. They must be greater than the many costs associated with exercising preferences. Arguably, this equation is different for choosers and courters. The question now is how this dynamic is altered when males also choose, or solely choose. Choosers, often females, have much to lose when they mate with suboptimal partners, in the worst case their whole lifetime reproductive success. The key costs for females or choosers result from sexual conflict (Johnstone and Keller, 2000). Assuming that multiple matings are more beneficial for males than females, this can lead to many forms of harassment where females have to employ strategies to cope with the male attention that goes beyond what is optimal for females. Such strategies can range from females allowing males to mate with them (convenience polyandry; Boulton et al., 2018) to aggressively confronting males. Harassment can also be costly because it detracts females from exercising behaviors that would be more beneficial to them, such as feeding (Plath et al., 2007; Makowicz and Schlupp, 2013). Generally, in addition to superfluous attention, mating can be costly by attracting attention from parties that are not part of the dyadic interaction between the two mating partners. This cost can be conspecifics watching, or parasites and predators exploiting the signals exchanged between the two partners. Most of that cost is borne by the courter, but because of the physical proximity to the courter, the chooser can also be affected (Wing, 1988). One well researched example is the tungara frog, *P. pustolosus*, where males call to attract females. The most efficient call, however, is also most likely to attract frog-eating bats (Hemingway et al., 2018) and parasitic flies (Legett et al., 2018). Similar phenomena are known for many mating systems.

Generally, there seems to be a tradeoff between attracting attention from potential mates and avoiding attention from eavesdroppers, including parasites and predators. Some species have evolved specific adaptations to this, such as "quiet song" in birds (Dabelsteen et al., 1998) or private channels that allow communication between mates, while excluding potential predators, parasites, and competitors (Cummings et al., 2003). Courters alone in many systems bear the cost of any premating activities and courting, especially if the choosers only spend very limited time with the courter, as is common in lek breeding systems (Rowe, 1994). It is not clear if actual matings really are much more costly than other activities, but it is indisputable that there are costs associated with the physical act of mating.

In addition, there are many more costs of a more general nature. For example, the time spent on mating, courtship, and mate searching cannot be spent on other activities, such as feeding or resting. There might be a cost in the form of cognitive load (e.g., as found for foraging decisions; Charalabidis et al., 2017; Hemingway et al., 2018), in the sense that mating may divert attention away from other behaviors. There is also opportunity cost, because while engaged in one mating another, better mating partner may go undetected. This, and cost associated with searching and evaluating of mating partners, is more associated with choosers. Furthermore, there is a risk and potentially actual cost from STDs. In many species, females can suffer physical damage from mating with males, but this is extreme in cases of traumatic insemination, where males inject their sperm directly into the female (Tatarnic et al., 2014). Such traumatic insemination has evolved several times in insects and at least once in spiders and is mainly found in Hemiptera (Tatarnic et al., 2014). In these species, males use needle-like copulatory organs to inseminate females into their body cavity. Traumatic insemination was first shown in bed bugs (*Cimex lectularius*) by Carayon in 1966 (Stutt and Siva-Jothy, 2001) and subsequently in several more species (Tatarnic et al., 2014).

All of these costs are well documented for choosers and should—at least in principle—also be relevant to choosing males. Some male costs of mating are associated with a more active male mating strategy, but many are independent of that. Unfortunately, not many studies seem to address the costs and benefits of male mate choice. In a theoretical paper, Heubel and colleagues investigated the evolution of male mate choice in the host species of the Amazon molly (*P. formosa*), a sexually parasitic species, and found that male mate choice did not evolve to perfectly recognizing conspecific and heterospecific females (Heubel et al., 2009). This means that allowing a number of matings with their sexual parasite is less costly for males of the host species *Poecilia latipinna* and *Poecilia mexicana* than evolving perfect recognition of conspecific females. Overall, the costs associated with female choice are fairly well understood, but costs associated with male choice are not.

2.15 Rejection as choice

Choosers typically have more potential mates than they are going to mate with. Consequently, they exercise mate choice not just by mating with the partner they prefer, but also by not mating with the males they do not prefer. Apparently, rejecting low-quality mates is—at least in simulations—different from choosing high-quality mates (Gomes and Cardoso, 2018). Furthermore, choice by rejection has a number of consequences, in which the rejected males try to circumvent female rejection. These male behaviors are usually costly to females and can affect female fitness negatively. Females, on the other hand can counter costly male behaviors in many ways ranging from accepting superfluous matings to fleeing or aggressively rejecting male efforts. This dynamic sets the stage for sexual conflict and evolution of female traits that are adaptive in countering attempts of male manipulation.

From the perspective of traditional sex roles, males are not expected to forgo many mating opportunities. But even for males, matings can be costly and if the costs exceed the benefits, males should decide not to mate. One important aspect of this is sperm depletion. Males may run out of sperm and, especially in cases of large ejaculates, males may need considerable time to replenish their supplies for new ejaculates. This means that females may benefit from choosing males that are not depleted (Hines et al., 2003), and that males may benefit from strategically using their resources (Preston et al., 2001), if sperm really is limited (Swierk et al., 2015). One predicted behavior is that males forgo matings with less attractive females. Male mate choice by rejection may be underreported, but some cases are known. In at least one case, males continue mating with females despite being sperm depleted and interestingly females were unable to detect this (Abe, 2019).

2.16 Social environment

Furthermore—and maybe most importantly—mate choice rarely happens in a social vacuum. In most species there are other individuals present for mate choice and the subsequent actual matings, which obviously applies to both females and males (Orfao

et al., 2019). This matters also, because the most common experimental technique to measure preferences uses an assay that involves a chooser and two alternative mates (or a representation of them, such as a video clip), but no audience. The advantage of using this approach is a controlled experimental design that should lead to repeatable data, but it lacks biological realism. As an aside: repeatability in behavioral data is an issue in itself, and sometimes difficult to obtain, but overall repeatability in the field of animal behavior is high (Bell et al., 2009). Low repeatability may be due to actual differences between studies but could also reflect suboptimal experimental design and data analysis. Viewing mate choice with others looking on in a broader social context points to a role of these onlookers in mating decisions (Agrillo et al., 2008). This has now been documented many times and we have many great examples of the dynamics of these social networks in which mating happens. In a scenario with onlookers both the focal individual and the onlooking individuals may modify their preferences and behaviors based on the interactions they experience. I want to revisit a relatively straightforward example for the importance of social information, mate choice copying (Nordell and Valone, 1998). As a concept first introduced in the 1980s (Losey et al., 1986; Höglund et al., 1990) and experimentally used in fishes (Dugatkin and Godin, 1992, 1993; Godin et al., 2005) in the 1990s and partly influenced by theory stemming from economics (Bikhchandani et al., 1992), the basic idea is simple: females are influenced by the mate preferences of other females. This can be a subtle modification of a preference or even a full reversal from a previously shown preference. Overall, effects of social information on mate choice are fairly strong (Jones and DuVal, 2019). Males also show mate copying, even in species where males are not very discriminatory otherwise (Schlupp and Ryan, 1997). In one case males were actually found to copy female mating decisions (Gonçalves et al., 2003). Some male pipefishes, where the sex roles are reversed, also show mate copying (Widemo, 2006), which in this case is the equivalent to female mate copying in other species. Interestingly, females (here the equivalent to males) do not show mate copying (Widemo, 2006).

In humans, mate copying can also be found (Waynforth, 2007; Anderson and Surbey, 2014; Deng and Zheng, 2015) and may be gender specific (Vakirtzis and Roberts, 2010). There is a large number of experimental studies, with somewhat mixed findings (Uller and Johansson, 2003), yet the consensus seems to be that mate copying in humans is common and an important mechanism in mate selection (Street et al., 2018), but there is an interesting sex difference that was revealed in a meta-analysis: mate copying is common in women, but not in men (Gouda-Vossos et al., 2018).

At least one example suggests that copying individuals are more likely to use information if it comes from a more reliable source. Female guppies are more likely to copy the preference of larger females, which are also older (Amlacher and Dugatkin, 2005). This is very suggestive; however, it does not establish a direct link between experience and usage of more reliable information. Another example could be that in a sexual–asexual mating complex involving the Amazon molly, *P. formosa*, sexual females are more likely to copy the mate preference of sexual females as compared with asexual females (Hill and Ryan, 2006).

More generally, the information provided in a social setting can be used by others and may provide valuable information for the observers. For example, males can obtain information about the behavior and the status of other males without directly interacting with them. This can save the cost of—for example—an escalated fight. This was found in males of the Siamese fighting fish (*Betta splendens*) watching other males fighting: they adjust their hormone levels according to likely future encounters (Oliveira et al., 1998, 2001). Another example comes from marmot's alarm calls—which evolved to be loud and detectable—but also can betray the parasite status of the caller (Nouri and Blumstein, 2019) to bystanders.

Another example of the role of the social environment is arranged marriages in humans. Historically, in many cultures mate choice was exercised or at least influenced by people other than the actual mating partners. Arranged marriages were common in many human societies (Davidson, 2012; Delameillieure, 2017) and are still widespread today (Hussain, 1984; Hampshire and Smith, 2001;

Al-Kandari et al., 2002; Williams and Pike, 2002). Arranged marriages are found, for example, in present-day Indian society (Sharangpani, 2010). Interestingly, it is still maintained to some degree in Indian immigrants in other countries, like the United Kingdom and the United States (Pichler, 2007). Clearly, arranged marriages are not restricted to a certain time or culture. Arranged marriages can also be highly exploitative and be connected to criminal human trafficking (Yakushko and Rajan, 2016).

In a bigger context it seems that the current Western model of free individual and rational choice (Ryan and Cummings, 2013), which we also apply to animals, is relatively new for humans on an evolutionary time scale (Coontz, 2006). In many societies—past and present—not the person to be married, but someone else decides about that relationship. Arranged relationships do undermine individual choice, but do not preclude a role of mate preferences in choice. What is exercised may not be the preference of the individual brought into a relationship, but that of older, more experienced individuals, often the parents (Lefevre and Saxton, 2017). This is somewhat similar to the horizontal mate copying discussed above. While the offspring may not be allowed to exercise their own preference, they have likely inherited that preference from their parents, and will be able to exercise the same preference when their children are getting married. There may be benefits from this system, as older individuals may be better able to assess the prospective partner regarding physical traits, health, etc. On the flip side, this creates strong parent–offspring conflict (van den Berg et al., 2013). However, a side effect of this may be a societal structure that leads to assortative mating, a pattern that has been found in many societies. In humans assortative mating often appears to be based on socio-economics, and that has maintained stratification for a long time. One example for this would be the caste system in India, where marriages between members of a different caste are at least frowned upon. Another example is arranged marriages in European royal houses, which tolerated the negative effects of inbreeding to maintain power within their lineage (Ceballos and Alvarez, 2013), although in the long run inbreeding may have contributed to their demise (Alvarez et al., 2009). This said, individual preferences may still be exercised and sexual interactions outside of the marriage can circumvent a choice made by others (Grebe and Drea, 2018).

In addition to directly determining mate choice in their children, human parents may influence more subtly who becomes their daughter-in-law or son-in-law by tacitly approving or disapproving of partners. Anecdotes and folklore about manipulative parents and in-laws are plenty and may indicate a deeply rooted conflict between generations about how to treat children and grandchildren. Cross-cultural research shows that parents value very similar characteristics in sons-in-law and daughters-in-law. The traits that were ranked highest for both were "Good character" and "Good family background." Physical traits summed up in "Good looking" were at the bottom of the list (Apostolou, 2007). This seems to be in contrast to the many studies on direct human mate choice that put more emphasis on physical traits like the hip to waist ratio, which was found to be almost universal. While clearly documented in humans, it is not clear whether such generational influences occur in other social animals. At least potentially they could occur in species where overlapping generations live together, and where parentage is known at least to some degree. Good candidates would be, for example, chimpanzees and bonobos, or orcas. In the latter species research into this question is ongoing (D. Croft, per. comm., September 12, 2018).

All of these effects of viewing mating in the proper social network are a powerful testament to the importance of social information in mate choice. Although much of this information is transmitted horizontally, some of that may become a component of nongenetic inheritance in the form of cultural evolution, which may be viewed as part of the extended evolutionary synthesis (Laland et al., 2015).

2.17 Similarities and differences

Many aspects of female and male choice are similar, especially on the proximate side, but how is male mate choice mirrored in the elements of female choice discussed above? Male and female choice can also differ in multiple ways. First, these two phenomena can have different evolutionary origins, and second, as outlined above, they may have different evolutionary consequences.

2.17.1 Evolutionary outcomes

Many species show differences between the sexes that go beyond the primary sex organs. Often males are under sexual selection and show traits that are interpreted as ornaments or armaments that evolved under either female choice or male competition. This view is supported by a huge literature and universally accepted. One might view the female phenotype almost as the default from which the male phenotype develops and evolves away. This view of course is incomplete as many different selection pressures operate on females. In raptors, for example, natural selection is thought to cause differences in size between females and males allowing them to exploit prey of different size. In this case it seems that natural selection may be mostly responsible for the dimorphism (Sonerud et al., 2014; Perez-Camacho et al., 2015) although some female preference for lighter males has been reported (Hakkarainen et al., 1996), whereas in many other cases natural and sexual selection interact.

In humans, it appears that many male characteristics are both armaments and ornaments (Berglund et al., 1996). They are likely to have evolved under sexual selection. Often traits may have evolved under male competition, but female choice is also invoked. Cleanly distinguishing between the two forces is often not possible and may not be necessary. Male traits associated with masculinity such as muscularity and deep voices are preferred by women and may be associated with male ability to forage or protect women (Buss and Schmitt, 2019). The same traits of course are adaptive in male–male competition. This is also found in many other species. If female choice influences the evolution of male traits, does male mate choice influence female traits? The answer seems to be yes: male preferences for traits associated with youth, indicating reproductive potential, such as reduced body hair and the distribution of body fat deposits, may be under sexual selection via male mate choice. The existence of body fat reserves is best explained by natural selection as body fat is necessary for successful ovulation and pregnancy. But the distribution of the deposits in breasts and on the hips is best explained by male mate choice (Puts, 2010). It is noteworthy that this distribution does not appear to give women an advantage in female–female competition or natural selection. Curiously in modern Western society, surgery techniques have emerged to alter female bodies supposedly to make them more attractive to men. Other traits in females may be interpreted as ornamental. The use of make-up may fall into this category, or alteration of the body via plastic surgery. However, while these phenomena may appear to agree with sexual selection theory they should be viewed with great caution. It is very likely that a large component of cultural influence and bias contributes to this view. We know that the ideal of beauty changes with culture and time period.

2.17.2 Mechanisms

Mate choice is not only influenced by external factors such as the social environment, there are many internal factors as well. Some of those have already been discussed, but it is important to remember that many physiological regulatory processes are relevant to mate choice. This ranges from overall regulating seasonality of mating in males and females, as well as small-scale adjustments to selectivity. At the interface of external and internal factors we find responses, for example, to starvation. In fruitflies, *D. melanogaster*, both female choosiness and male courtship vigor were reduced in hungry flies (Billings et al., 2018). Looking at the bigger picture this raises the question of whether some studies on mate choice that were conducted on satiated females and males might have given us an incomplete picture.

One especially interesting aspect we need to consider is that both sexes may be evaluating very different traits when they exercise preferences. For humans this was tackled in a meta-analysis of preferences (Conroy-Beam et al., 2015). The authors found that the two genders they recognize differ in major ways in what they evaluate in mate choice. In other words, humans are very dimorphic in their mating preferences (Buss and Schmitt, 2019). The

authors point out that—as with all studies on humans—some caution in interpreting the findings is advised because of potential cultural influences. Nonetheless, the message is clear: typical men and typical women are attracted to very different traits in the opposite sex. Clearly, an evaluation of same-sex attraction would be very valuable in this context. Complicating things, both sexes may have the same preference for a trait, but only apparently so. One example is the widespread preference for larger size, which may be beneficial as a direct benefit in males and an indirect benefit in females (Komdeur et al., 2005; Schlupp, 2018). The argument is simple: in many females, the number of eggs or embryos is positively correlated with size. Larger females have a higher fecundity and male preference for larger females is easily explained. Males that successfully mate with a more fecund female have more offspring. The argument for female preferences often typically involves indirect benefits. In many species, males contribute nothing to the offspring other than sperm. In these cases, female preference for larger males cannot be explained via direct benefits. At best, there may be weak effects because—for example—larger males produce more sperm. But in essence, females preferring larger males must gain an indirect benefit, potentially because their large sons will also be preferred by females. The literature on such preferences is overwhelming in reporting such female preferences for larger males. It is not clear if this might be a case of publication bias, where studies not providing a confirmation of previous results have a higher probability of being rejected. Indeed, there are very few published studies of a lack of female preference for large males. These studies do show that an absence of a preference can be documented, but they also have a hard time interpreting their findings. Clearly, however, males and females differ fundamentally in the likely evolutionary benefit of a preference for large body size. Maybe the widespread female preference is a pleiotropic effect of the male preference for female fecundity.

It must be said here that in many species preferring larger males is providing direct benefits for females. Furthermore, larger male body size is advantageous in many aspects of male competition, such as mate guarding, nest guarding, provisioning of mates and offspring, defense of mate and off-spring, and most prominently in male fighting (Tinbergen, 1936; Parker, 1974; Davies and Halliday, 1978; Clutton-Brock et al., 1989; Allen et al., 2017). Under both scenarios, natural and sexual selection sexual dimorphism will increase. One could argue that competition for mates has recently received less attention than choice of mates, but its important role in sexual selection is clear.

How mate choice is exercised is also important, and I want to briefly summarize a few aspects, including limitations of mate choice. Western societies, where the majority of mate choice studies are conducted, emphasize freedom in choosing a mating partner, but this appears to be a relatively new development, and is not universal across human cultures.

Arranged marriages also provide an example of social influences on mate choice found in different forms in many other animals (as discussed above). These social influences are important in several ways. First, when studied experimentally, mate choice is often reduced to an interaction between only two individuals. In reality, though, mate choice is exercised in public, often with several onlookers, who may extract information about the mating partners from the interaction. The onlookers may then adjust their own behavior in several ways. Both female and male choice are thus exposed to various audiences. One particular effect is called mate copying and refers to females and males observing the mate preferences of others and changing their own mating behavior accordingly. This has been observed in many species now, including humans.

2.18 Hermaphrodites

So far, we have considered organisms where the sexes are located in different individuals. This is not always the case and many species are hermaphrodites with both sexes present in the same individuals either at the same time or after a sex change (Charnov et al., 1976). The taxonomic distribution is biased toward invertebrates, with very few vertebrates having evolved hermaphroditism, but in general this is a very common situation. The number of hermaphroditic animal species was estimated

to be 65,000 (Jarne and Auld, 2006), which is a very small fraction of all species. In higher plants a comparable situation is found where the sexual phase (the gametophyte) is monoicous, meaning that the same individual plant carries male and female gametes. For humans the term hermaphrodite is also used in the medical literature, sometimes interchangeable with "intersex" and in the context of human sex reassignment (Besnier, 2018). My discussion here is focusing on nonhuman animals.

Theory would predict that sequential hermaphrodites express one sex role while male and another while female. Quite a few species show such sequential hermaphroditism, where an individual may first be male, then female (protandry) or first female and then male (protogyny). This raises the interesting question of consistency of preferences. Suppose an animal is first female and shows a preference for larger males. It then changes into a male, as found in many wrasses, but what happens to the preference?

Sequential hermaphrodites also provide an interesting perspective on the economy of sex and sex roles. In these species the costs and benefits of having one sex can be compared with the other sex and animals change sex when the two curves intersect. It is ecology that influences that decision.

In addition to natural sex change, sometimes hormones (and their mimics) in the environment can induce sex change in species that are normally bimodal in their sex expression (Bhandari et al., 2015). This gives important clues as to the mechanisms underlying sexual development involving steroid hormones (Nozu et al., 2009), which are also crucially important under natural conditions (Frisch, 2004). Mainly in insects, the bacterial parasite *Wolbachia* is also causing feminization (Werren et al., 2008).

In simultaneous hermaphrodites, because female and male behaviors must be present in the same organism at the same time, a number of very interesting questions arise regarding the regulation of interactions. This has been extensively modeled and reported in a number of studies starting with seminal work by Charnov in 1979 (Charnov, 1979; Shifferman, 2012; Jordan and Connallon, 2014; Parker, 2016; Parma et al., 2016; Koene, 2017; Larson, 2017). One interesting problem is that the male component of a simultaneous hermaphrodite may be interested in inseminating another individual, while the female part may not be interested in receiving sperm. Clearly, mating decisions made by the male component of the individual will impact directly the decisions available to the female component. Interestingly, sperm trading can narrow the gap in Bateman gradients in hermaphrodites, leading to essentially equal investment in male and female components (Greeff and Michiels, 1999). Also, male–male competition has led to behaviors like sperm fencing and forced insemination (Michiels, 1998). A review is provided by Beekman et al. (2016). Added complexity results from the fact that some hermaphrodites are capable of selfing (fertilizing their own eggs with their sperm) and that some species have both hermaphrodites and non-hermaphroditic males (Weeks et al., 2006; Berbel-Filho et al., 2020) or females (Dufay and Billard, 2012). In a typical binary choice test based on associations, Lüscher and Wedekind found a preference for large size in the parasitic worm *Schistocephalus solidus* (Lüscher and Wedekind, 2002).

2.19 Short summary

In this chapter I discussed the conditions under which female and male choice may evolve and the theory behind this. I further discussed the key features of and main differences between female and male mate choice. While female choice seems to evolve easily based on the ecology of choice and the fact that females often take little to no risk when they forgo a particular mating opportunity, male choice is less likely to evolve because mating opportunities tend be rarer for males and the cost of missing a particular mating opportunity can be so high that male choice is unlikely to evolve.

2.20 Recommended reading

The best book to read to continue thinking about mate choice more generally is Rosenthal (2017). I also think that reading the classic book on the topic by Andersson is still a good idea (Andersson, 1994). For a good overview of male mate choice I recommend Edward and Chapman (2011) and Bonduriansky (2001).

2.21 References

ABE, J. 2019. Sperm-limited males continue to mate, but females cannot detect the male state in a parasitoid wasp. *Behavioral Ecology and Sociobiology,* 73, 52.

ACHORN, A.M. & ROSENTHAL, G.G. 2020. It's not about him: mismeasuring "good genes" in sexual selection. *Trends in Ecology & Evolution,* 35, 206–19.

AGRILLO, C., DADDA, M., & SERENA, G. 2008. Choice of female groups by male mosquitofish (*Gambusia holbrooki*). *Ethology,* 114, 479–88.

AH-KING, M. & AHNESJÖ, I. 2013. The "sex role" concept: an overview and evaluation. *Evolutionary Biology,* 40, 461–70.

AHNESJÖ, I., BREALEY, J. C., GÜNTER, K. P., MARTINOSSI-ALLIBERT, I., MORINAY, J., SILJESTAM, M., STÅNGBERG, J., & VASCONCELOS, P. 2020. Considering gender-biased assumptions in evolutionary biology. *Evolutionary Biology,* 47, 1–5.

AL-KANDARI, Y., CREWS, D. E., & POIRIER, F. E. 2002. Length of marriage and its effect on spousal concordance in Kuwait. *American Journal of Human Biology,* 14, 1–8.

ALEXANDER, R. D. & OTTE, D. 1967. Cannibalism during copulation in the brown bush cricket, *Hapithus agitator* (Gryllidae). *The Florida Entomologist,* 50, 79–87.

ALLEN, J. J., AKKAYNAK, D., SCHNELL, A. K., & HANLON, R. T. 2017. Dramatic fighting by male cuttlefish for a female mate. *The American Naturalist,* 190, 144–51.

ALVAREZ, G., CEBALLOS, F. C., & QUINTEIRO, C. 2009. The role of inbreeding in the extinction of a European royal dynasty. *PLoS ONE,* 4, e5174.

AMLACHER, J. & DUGATKIN, L. A. 2005. Preference for older over younger models during mate-choice copying in young guppies. *Ethology Ecology & Evolution,* 17, 161–9.

AMUNDSEN, T. 2018. Sex roles and sexual selection: lessons from a dynamic model system. *Current Zoology,* 64, 363–92.

ANCONA, S., DÉNES, F. V., KRÜGER, O., SZÉKELY, T., & BEISSINGER, S. R. 2017. Estimating adult sex ratios in nature. *Philosophical Transactions of the Royal Society B—Biological Sciences,* 372, 20160313.

ANDERSON, R. C. & SURBEY, M. K. 2014. "I want what she's having": evidence of human mate copying. *Human Nature—an Interdisciplinary Biosocial Perspective,* 25, 342–58.

ANDERSSON, M. 1994. *Sexual Selection.* Princeton, NJ: Princeton University Press.

APOSTOLOU, M. 2007. Sexual selection under parental choice: the role of parents in the evolution of human mating. *Evolution and Human Behavior,* 28, 403–9.

ARNOLD, S. J. 1994. Bateman's principles and the measurement of sexual selection in plants and animals. *The American Naturalist*, 144, S126–49.

BABA, N., HIRONAKA, M., HOSOKAWA, T., MUKAI, H., NOMAKUCHI, S., & UENO, T. 2011. Trophic eggs compensate for poor offspring feeding capacity in a subsocial burrower bug. *Biology Letters*, 7, 194–6.

BALSHINE-EARN, S. & EARN, D. J. 1998. On the evolutionary pathway of parental care in mouth-brooding cichlid fishes. *Proceedings of the Royal Society of London— Series B: Biological Sciences*, 265, 2217–22.

BARATTI, M., DESSI-FULGHERI, F., AMBROSINI, R., BONISOLI-ALQUATI, A., CAPRIOLI, M., GOTI, E., MATTEO, A., MONNANNI, R., RAGIONIERI, L., RISTORI, E., ROMANO, M., RUBOLINI, D., SCIALPI, A., & SAINO, N. 2012. MHC genotype predicts mate choice in the ring-necked pheasant *Phasianus colchicus*. *Journal of Evolutionary Biology*, 25, 1531–42.

BASOLO, A. L. & ENDLER, J. A. 1995. Sensory biases and the evolution of sensory systems. *Trends in Ecology & Evolution*, 10, 489–9.

BATEMAN, A. J. 1948. Intra-sexual selection in *Drosophila*. *Heredity*, 2, 349–68.

BEECHER, M. D. & BRENOWITZ, E. A. 2005. Functional aspects of song learning in songbirds. *Trends in Ecology & Evolution*, 20, 143–9.

BEEKMAN, M., NIEUWENHUIS, B., ORTIZ-BARRIENTOS, D., & EVANS, J. P. 2016. Sexual selection in hermaphrodites, sperm and broadcast spawners, plants and fungi. *Philosophical Transactions of the Royal Society B—Biological Sciences*, 371. https://doi.org/10.1098/rstb.2015.0541.

BELL, A. M., HANKISON, S. J., & LASKOWSKI, K. L. 2009. The repeatability of behaviour: a meta-analysis. *Animal Behaviour*, 77, 771–83.

BERBEL-FILHO, W. M., TATARENKOV, A., ESPÍRITO-SANTO, H. M. V., LIRA, M. G., GARCIA DE LEANIZ, C., LIMA, S. M. Q., & CONSUEGRA, S. 2020, 125, 340–352. More than meets the eye: syntopic and morphologically similar mangrove killifish species show different mating systems and patterns of genetic structure along the Brazilian coast. *Heredity*, 125, 340–52.

BERGLUND, A., BISAZZA, A., & PILASTRO, A. 1996. Armaments and ornaments: an evolutionary explanation of traits of dual utility. *Biological Journal of the Linnean Society*, 58, 385–99.

BERGLUND, A., MAGNHAGEN, C., BISAZZA, A., KONIG, B., & HUNTINGFORD, F. 1993. Female–female competition over reproduction. *Behavioral Ecology*, 4, 184–7.

BERGLUND, A. & ROSENQVIST, G. 2001. Male pipefish prefer ornamented females. *Animal Behaviour*, 61, 345–50.

BERGLUND, A. & ROSENQVIST, G. 2003. Sex role reversal in pipefish. *In*: SLATER, P. J. B., ROSENBLATT, J. S., SNOWDON, C. T., & ROPER, T. J. (eds.) *Advances in the Study of Behavior, Vol. 32*. Amsterdam: Elsevier.

BERGLUND, A., ROSENQVIST, G., & BERNET, P. 1997. Ornamentation predicts reproductive success in female pipefish. *Behavioral Ecology and Sociobiology*, 40, 145–50.

BERGLUND, A., ROSENQVIST, G., & SVENSSON, I. 1986. Mate choice, fecundity and sexual dimorphism in two pipefish species (Syngnathidae). *Behavioral Ecology and Sociobiology*, 19, 301–7.

BESNIER, N. 2018. Sex reassignment. *The International Encyclopedia of Anthropology*, 1–2. https://doi.org/10.1002/9781118924396.wbiea1360.

BHANDARI, R. K., DEEM, S. L., HOLLIDAY, D. K., JANDEGIAN, C. M., KASSOTIS, C. D., NAGEL, S. C., TILLITT, D. E., VOM SAAL, F. S., & ROSENFELD, C. S. 2015. Effects of the environmental estrogenic contaminants bisphenol A and 17α-ethinyl estradiol on sexual development and adult behaviors in aquatic wildlife species. *General and Comparative Endocrinology*, 214, 195–219.

BIKHCHANDANI, S., HIRSHLEIFER, D., & WELCH, I. 1992. A theory of fads, fashion, custom, and cultural change as informational cascades. *Journal of Political Economy*, 100, 992–1925.

BILLIARD, S., LÓPEZ-VILLAVICENCIO, M., DEVIER, B., HOOD, M. E., FAIRHEAD, C., & GIRAUD, T. 2011. Having sex, yes, but with whom? Inferences from fungi on the evolution of anisogamy and mating types. *Biological Reviews*, 86, 421–42.

BILLIARD, S., LÓPEZ-VILLAVICENCIO, M., HOOD, M., & GIRAUD, T. 2012. Sex, outcrossing and mating types: unsolved questions in fungi and beyond. *Journal of Evolutionary Biology*, 25, 1020–38.

BILLINGS, A. C., SCHULTZ, K. E., HERNANDEZ, E. A., JONES, W. E. & PRICE, D. K. 2018. Male courtship behaviors and female choice reduced during experimental starvation stress. *Behavioral Ecology*, 30, 231–9.

BJORK, A. & PITNICK, S. 2006. Intensity of sexual selection along the anisogamy–isogamy continuum. *Nature*, 441, 742–5.

BONDURIANSKY, R. 2001. The evolution of male mate choice in insects: a synthesis of ideas and evidence. *Biological Reviews*, 76, 305–39.

BONDURIANSKY, R. 2009. Reappraising sexual coevolution and the sex roles. *PLoS Biology*, 7, e1000255.

BOOGERT, N. J., LACHLAN, R. F., SPENCER, K. A., TEMPLETON, C. N., & FARINE, D. R. 2018. Stress hormones, social associations and song learning in zebra finches. *Philosophical Transactions of the Royal Society B—Biological Sciences*, 373, 20170290.

BOULTON, R. A., ZUK, M., & SHUKER, D. M. 2018. An inconvenient truth: the unconsidered benefits of convenience polyandry. *Trends in Ecology & Evolution*, 33, 904–15.

BRADBURY, J. W. & VEHRENCAMP, S. 2011. *Principles of Animal Communication*. Sunderland, MA: Sinauer.

BRAGA GONÇALVES, I., MOBLEY, K. B., AHNESJO, I., SAGEBAKKEN, G., JONES, A. G., & KVARNEMO, C. 2015. Effects of mating order and male size on embryo survival in a pipefish. *Biological Journal of the Linnean Society*, 114, 639–45.

BRENOWITZ, E. A. & BEECHER, M. D. 2005. Song learning in birds: diversity and plasticity, opportunities and challenges. *Trends in Neurosciences*, 28, 127–32.

BURGER, D., MEUWLY, C., MARTI, E., SIEME, H., OBERTHUR, M., JANDA, J., MEINECKE-TILLMANN, S., & WEDEKIND, C. 2017. MHC-correlated preferences in diestrous female horses (*Equus caballus*). *Theriogenology*, 89, 318–23.

BUSS, D. M. 1989. Sex differences in human mate preferences: evolutionary hypotheses tested in 37 cultures. *Behavioral and Brain Sciences*, 12, 1–14.

BUSS, D. M. & SCHMITT, D. P. 2019. Mate preferences and their behavioral manifestations. *Annual Review of Psychology*, 70, 77–110.

BUTCHART, S. H. M. 2000. Population structure and breeding system of the sex-role reversed, polyandrous bronze-winged jacana *Metopidius indicus. Ibis*, 142, 93–102.

CARLETON, K. L., PARRY, J. W. L., BOWMAKER, J. K., HUNT, D. M., & SEEHAUSEN, O. 2005. Colour vision and speciation in Lake Victoria cichlids of the genus *Pundamilia. Molecular Ecology*, 14, 4341–53.

CARMONA-ISUNZA, M. C., ANCONA, S., SZEKELY, T., RAMALLO-GONZALEZ, A. P., CRUZ-LOPEZ, M., SERRANO-MENESES, M. A., & KUPPER, C. 2017. Adult sex ratio and operational sex ratio exhibit different temporal dynamics in the wild. *Behavioral Ecology*, 28, 523–32.

CEBALLOS, F. C. & ALVAREZ, G. 2013. Royal dynasties as human inbreeding laboratories: the Habsburgs. *Heredity*, 111, 114–21.

CHANG, C.-H. & YAN, H. Y. 2019. Plasticity of opsin gene expression in the adult red shiner (*Cyprinella lutrensis*) in response to turbid habitats. *PLoS ONE*, 14, e0215376.

CHAPMAN, T., ARNQVIST, G., BANGHAM, J., & ROWE, L. 2003. Sexual conflict. *Trends in Ecology & Evolution*, 18, 41–7.

CHAPMAN, T., LIDDLE, L. F., KALB, J. M., WOLFNER, M. F., & PARTRIDGE, L. 1995. Cost of mating in *Drosophila melanogaster* females is mediated by male accessory gland products. *Nature*, 373, 241.

CHARALABIDIS, A., DECHAUME-MONCHARMONT, F. X., PETIT, S., & BOHAN, D. A. 2017. Risk of predation makes foragers less choosy about their food. *PLoS*, 12, e0187167.

CHARNOV, E. L. 1979. Simultaneous hermaphroditism and sexual selection. *Proceedings of the National Academy of Sciences*, 76, 2480–4.

CHARNOV, E. L., BULL, J. J., & MAYNARD SMITH, J. 1976. Why be an hermaphrodite? *Nature*, 263, 125–6.

CLUTTON-BROCK, T. H., HIRAIWA-HASEGAWA, M., & ROBERTSON, A. 1989. Mate choice on fallow deer leks. *Nature*, 340, 463–5.

COLLET, J. M., DEAN, R. F., WORLEY, K., RICHARDSON, D. S., & PIZZARI, T. 2014. The measure and significance of Bateman's principles. *Proceedings of the Royal Society of London—Series B: Biological Sciences*, 281, p.20132973.

COLWELL, M. A. & ORING, L. W. 1988. Sex-ratios and intrasexual competition for mates in a sex-role reversed shorebird, Wilson phalarope (*Phalaropus tricolor*). *Behavioral Ecology and Sociobiology*, 22, 165–73.

CONROY-BEAM, D., BUSS, D. M., PHAM, M. N., & SHACKELFORD, T. K. 2015. How sexually dimorphic are human mate preferences? *Personality and Social Psychology Bulletin*, 41, 1082–93.

COONTZ, S. 2006. *Marriage, a History: How Love Conquered Marriage*. London: Penguin.

CREAN, A. J., ADLER, M. I., & BONDURIANSKY, R. 2016. Seminal fluid and mate choice: new predictions. *Trends in Ecology & Evolution*, 31, 253–5.

CUI, R., SCHUMER, M., KRUESI, K., WALTER, R., ANDOLFATTO, P., & ROSENTHAL, G. G. 2013. Phylogenomics reveals extensive reticulate evolution in *Xiphophorus* fishes. *Evolution*, 67, 2166–79.

CUMMINGS, M. E., ROSENTHAL, G. G., & RYAN, M. J. 2003. A private ultraviolet channel in visual communication. *Proceedings of the Royal Society of London—Series B: Biological Sciences*, 270, 897–904.

DABELSTEEN, T., MCGREGOR, P. K., LAMPE, H. M., LANGMORE, N. E., & HOLLAND, J. 1998. Quiet song in song birds: an overlooked phenomenon. *Bioacoustics*, 9, 89–105.

DANCHIN, E., GIRALDEAU, L.-A., VALONE, T. J., & WAGNER, R. H. 2004. Public information: from nosy neighbors to cultural evolution. *Science*, 305, 487–91.

DAVIDSON, D. Z. 2012. "Happy" marriages in early nineteenth-century France. *Journal of Family History*, 37, 23–35.

DAVIES, N. B. & HALLIDAY, T. R. 1978. Deep croaks and fighting assessment in toads *Bufo bufo. Nature*, 274, 683.

DAWKINS, R. & KREBS, J. R. 1979. Arms races between and within species. *Proceedings of the Royal Society London B—Biological Sciences*, 205, 489–511.

DAWLEY, R. M. 1989. An introduction to unisexual vertebrates. *In*: DAWLEY, R. M. & BOGART, J. P. (eds.) *Evolution and Ecology of Unisexual Vertebrates*, Bulletin 466. Albany, NY: New York State Museum.

DE QUEIROZ, K. 2005. Ernst Mayr and the modern concept of species. *Proceedings of the National Academy of Sciences*, 102, 6600–7.

DELAMEILLIEURE, C. 2017. "Partly with and partly against her will": female consent, elopement, and abduction in late medieval Brabant. *Journal of Family History*, 42, 351–68.

DENG, Y. & ZHENG, Y. 2015. Mate-choice copying in single and coupled women: the influence of mate acceptance and mate rejection decisions of other women. *Evolutionary Psychology*, 13, 89–105.

DOTY, G. V. & WELCH, A. M. 2001. Advertisement call duration indicates good genes for offspring feeding rate in gray tree frogs (*Hyla versicolor*). *Behavioral Ecology and Sociobiology*, 49, 150–6.

DUFAY, M. & BILLARD, E. 2012. How much better are females? The occurrence of female advantage, its proximal causes and its variation within and among gynodioecious species. *Annals of Botany*, 109, 505–19.

DUGAS, M. B., WAMELINK, C. N., KILLIUS, A. M., & RICHARDS-ZAWACKI, C. L. 2016. Parental care is beneficial for offspring, costly for mothers, and limited by family size in an egg-feeding frog. *Behavioral Ecology*, 27, 476–83.

DUGATKIN, L. A. & GODIN, J.-G. J. 1992. Reversal of female mate choice by copying in the guppy *Poecilia reticulata*. *Proceedings of the Royal Society of London—Series B: Biological Sciences*, 249, 179–84.

DUGATKIN, L. A. & GODIN, J.-G. J. 1993. Female mate copying in the guppy (*Poecilia reticulata*): Age-dependent effects. *Behavioral Ecology*, 4, 289–92.

EDELMAN, N. B., FRANDSEN, P. B., MIYAGI, M., CLAVIJO, B., DAVEY, J., DIKOW, R. B., GARCÍA-ACCINELLI, G., VAN BELLEGHEM, S. M., PATTERSON, N., & NEAFSEY, D. E. 2019. Genomic architecture and introgression shape a butterfly radiation. *Science*, 366, 594–9.

EDWARD, D. A. & CHAPMAN, T. 2011. The evolution and significance of male mate choice. *Trends in Ecology & Evolution*, 26, 647–54.

EENS, M. & PINXTEN, R. 2000. Sex-role reversal in vertebrates: behavioural and endocrinological accounts. *Behavioural Processes*, 51, 135–47.

EMLEN, S. T., DEMONG, N. J., & EMLEN, D. J. 1989. Experimental induction of infanticide in female wattled jacanas. *Auk*, 106, 1–7.

EMLEN, S. T. & ORING, L. W. 1977. Ecology, sexual selection, and the evolution of mating systems. *Science*, 197, 215–23.

EMLEN, S. T. & WREGE, P. H. 2004. Size dimorphism, intrasexual competition, and sexual selection in wattled jacana (*Jacana jacana*), a sex-role-reversed shorebird in Panama. *Auk*, 121, 391–403.

ENGQVIST, L. 2007. Sex, food and conflicts: nutrition dependent nuptial feeding and pre-mating struggles in scorpionflies. *Behavioral Ecology and Sociobiology*, 61, 703–10.

FITZPATRICK, C. L. & SERVEDIO, M. R. 2017. Male mate choice, male quality, and the potential for sexual selection on female traits under polygyny. *Evolution*, 71, 174–83.

FITZPATRICK, C. L. & SERVEDIO, M. R. 2018. The evolution of male mate choice and female ornamentation: a review of mathematical models. *Current Zoology*, 64, 323–33.

FORSGREN, E., AMUNDSEN, T., BORG, Å. A., & BJELVENMARK, J. 2004. Unusually dynamic sex roles in a fish. *Nature*, 429, 551.

FRISCH, A. 2004. Sex-change and gonadal steroids in sequentially-hermaphroditic teleost fish. *Reviews in Fish Biology and Fisheries*, 14, 481–99.

FROMHAGE, L., UHL, G., & SCHNEIDER, J. M. 2003. Fitness consequences of sexual cannibalism in female *Argiope bruennichi*. *Behavioural Ecology and Sociobiology*, 55, 60–4.

GABOR, C. R. & RYAN, M. J. 2001. Geographical variation in reproductive character displacement in mate choice by male sailfin mollies. *Proceedings of the Royal Society of London—Series B: Biological Sciences*, 268, 1063–70.

GABOR, C. R., RYAN, M. J., & MORIZOT, D. C. 2005. Character displacement in sailfin mollies, *Poecilia latipinna*: allozymes and behavior. *Environmental Biology of Fishes*, 73, 75–88.

GIBSON, R. & BAKER, A. 2012. Multiple gene sequences resolve phylogenetic relationships in the shorebird suborder Scolopaci (Aves: Charadriiformes). *Molecular Phylogenetics and Evolution*, 64, 66–72.

GODIN, J. G. J., HERDMAN, E. J. E., & DUGATKIN, E. A. 2005. Social influences on female mate choice in the guppy, *Poecilia reticulata*: generalized and repeatable trait-copying behaviour. *Animal Behaviour*, 69, 999–1005.

GOMES, A. C. R. & CARDOSO, G. C. 2018. Choice of high-quality mates versus avoidance of low-quality mates. *Evolution*, 72, 2608–16.

GONÇALVES, D., OLIVEIRA, R. F., KORNER, K., & SCHLUPP, I. 2003. Intersexual copying by sneaker males of the peacock blenny. *Animal Behaviour*, 65, 355–61.

GOODWIN, N. B., BALSHINE-EARN, S., & REYNOLDS, J. D. 1998. Evolutionary transitions in parental care in cichlid fish. *Proceedings of the Royal Society of London—Series B: Biological Sciences*, 265, 2265–72.

GOUDA-VOSSOS, A., NAKAGAWA, S., DIXSON, B. J. W., & BROOKS, R. C. 2018. Mate choice copying in humans: a systematic review and meta-analysis. *Adaptive Human Behavior and Physiology*, 4, 364–86.

GRANT, P. R. & GRANT, B. R. 1992. Hybridization of bird species. *Science*, 256, 193–7.

GRANT, P. R. & GRANT, B. R. 2019. Adult sex ratio influences mate choice in Darwin's finches. *Proceedings of the National Academy of Sciences*, 116, 12373–82.

GREBE, N. M. & DREA, C. M. 2018. Human sexuality. *In*: SHACKELFORD, T. K. & WEEKES-SHACKELFORD, V. A. (eds.) *Encyclopedia of Evolutionary Psychological Science*. Gland, Switzerland: Springer.

GREEFF, J. M. & MICHIELS, N. K. 1999. Sperm digestion and reciprocal sperm transfer can drive hermaphrodite sex allocation to equality. *The American Naturalist*, 153, 421–30.

GREEN, D. M. 2019. Rarity of size-assortative mating in animals: assessing the evidence with anuran amphibians. *The American Naturalist*, 193, 279–95.

GREENFIELD, M. D., ALEM, S., LIMOUSIN, D., & BAILEY, N. W. 2014. The dilemma of Fisherian sexual selection: mate choice for indirect benefits despite rarity and overall weakness of trait-preference genetic correlation. *Evolution*, 68, 3524–36.

GWYNNE, D. T. 1986. Courtship feeding in katydids (Orthoptera, Tettigoniidae)—investment in offspring or in obtaining fertilizations. *American Naturalist*, 128, 342–52.

HAKKARAINEN, H., HUHTA, E., LAHTI, K., LUNDVALL, P., MAPPES, T., TOLONEN, P., & WIEHN, J. 1996. A test of male mating and hunting success in the kestrel: the advantages of smallness? *Behavioral Ecology and Sociobiology*, 39, 375–80.

HAMPSHIRE, K. R. & SMITH, M. T. 2001. Consanguineous marriage among the Fulani. *Human Biology*, 73, 597–603.

HAVLICEK, J. & ROBERTS, S. C. 2009. MHC-correlated mate choice in humans: a review. *Psychoneuroendocrinology*, 34, 497–512.

HEMINGWAY, C. T., RYAN, M. J., & PAGE, R. A. 2018. Cognitive constraints on optimal foraging in frog-eating bats. *Animal Behaviour*, 143, 43–50.

HEMINGWAY, C. T., RYAN, M. J., & PAGE, R. A. 2019. Transitive foraging behaviour in frog-eating bats. *Animal Behaviour*, 154, 47–55.

HENSHAW, J. M., KAHN, A. T., & FRITZSCHE, K. 2016. A rigorous comparison of sexual selection indexes via simulations of diverse mating systems. *Proceedings of the National Academy of Sciences of the United States of America*, 113, E300–8.

HEUBEL, K. 2018. Female mating competition alters female mating preferences in common gobies. *Current Zoology*, 64, 351–61.

HEUBEL, K. U., HORNHARDT, K., OLLMANN, T., PARZEFALL, J., RYAN, M. J., & SCHLUPP, I. 2008. Geographic variation in female mate-copying in the species complex of a unisexual fish, *Poecilia formosa*. *Behaviour*, 145, 1041–64.

HEUBEL, K. U., RANKIN, D. J., & KOKKO, H. 2009. How to go extinct by mating too much: population consequences of male mate choice and efficiency in a sexual-asexual species complex. *Oikos*, 118, 513–20.

HILL, S. E. & RYAN, M. J. 2006. The role of model female quality in the mate choice copying behaviour of sailfin mollies. *Biology Letters*, 2, 203–5.

HINES, A. H., JIVOFF, P. R., BUSHMANN, P. J., VAN MONTFRANS, J., REED, S. A., WOLCOTT, D. L., & WOLCOTT, T. G. 2003. Evidence for sperm limitation in the blue crab, *Callinectes sapidus*. *Bulletin of Marine Science*, 72, 287–310.

HÖGLUND, J., ALATALO, R. V., & LUNDBERG, A. 1990. Copying the mate choice of others—observations on female black grouse. *Behaviour*, 114, 221–31.

HÖHN, E. O. 1965. *Die Wassertreter*. Leipzig, Germany: A. Ziemsen.

HOKE, K. L., RYAN, M. J., & WILCZYNSKI, W. 2010. Sexually dimorphic sensory gating drives behavioral differences in tungara frogs. *Journal of Experimental Biology*, 213, 3463.

HOLLAND, B. & RICE, W. R. 1999. Experimental removal of sexual selection reverses intersexual antagonistic coevolution and removes a reproductive load. *Proceedings of the National Academy of Sciences*, 96, 5083–8.

HOLWELL, G. I., WINNICK, C., TREGENZA, T., & HERBERSTEIN, M. E. 2010. Genital shape correlates with sperm transfer success in the praying mantis *Ciulfina klassi* (Insecta: Mantodea). *Behavioral Ecology and Sociobiology*, 64, 617–25.

HOOVER, B. & NEVITT, G. 2016. Modeling the importance of sample size in relation to error in MHC-based mate-choice studies on natural populations. *Integrative and Comparative Biology*, 56, 925–33.

HOQUET, T. 2020. Bateman (1948): rise and fall of a paradigm? *Animal Behaviour*, 164, 223–31.

HUBBS, C. 1964. Interactions between bisexual fish species and its gynogenetic sexual parasite. *Bulletin of the Texas Memorial Museum*, 8, 1–72.

HUBBS, C. L. & HUBBS, L. C. 1932. Apparent parthenogenesis in nature in a form of fish of hybrid origin. *Science*, 76, 628–30.

HUSSAIN, A. 1984. Assortative mating in arranged marriages—a study of a Muslim population in Hyderabad, Andhra Pradesh. *Annals of Human Biology*, 11, 472–2.

HUSTLER, K. & DEAN, W. R. J. 2002. Observations on the breeding biology and behaviour of the lesser jacana, *Microparra capensis*. *Ostrich*, 73, 79–82.

JANICKE, T., HAEDERER, I. K., LAJEUNESSE, M. J., & ANTHES, N. 2016. Darwinian sex roles confirmed across the animal kingdom. *Science Advances*, 2.

JARNE, P. & AULD, J. R. 2006. Animals mix it up too: the distribution of self-fertilization among hermaphroditic animals. *Evolution*, 60, 1816–24.

JAWORSKA, J., RAPACZ-LEONARD, A., JANOWSKI, T. E., & JEZIERSKI, T. 2017. Does the major histocompatibility complex influence choice of mate in humans and other mammals?—a review. *Animal Science Papers and Reports*, 35, 107–21.

JIANG, Y., BOLNICK, D. I., & KIRKPATRICK, M. 2013. Assortative mating in animals. *The American Naturalist*, 181, E125–38.

JOHNSTONE, R. A. & KELLER, L. 2000. How males can gain by harming their mates: sexual conflict, seminal toxins, and the cost of mating. *The American Naturalist*, 156, 368–77.

JOKINIEMI, A., MAGRIS, M., RITARI, J., KUUSIPALO, L., LUNDGREN, T., PARTANEN, J., & KEKÄLÄINEN, J. 2020. Post-copulatory genetic matchmaking: HLA-dependent effects of cervical mucus on human sperm function. *Proceedings of the Royal Society of London—Series B: Biological Sciences*, 287, 20201682.

JONASON, P. K., BETES, S. L., & LI, N. P. 2019. Solving mate shortages: lowering standards, searching farther, and abstaining. *Evolutionary Behavioral Sciences*, 14, 160.

JONES, A. G. 2009. On the opportunity for sexual selection, the Bateman gradient and the maximum intensity of sexual selection. *Evolution: International Journal of Organic Evolution*, 63, 1673–84.

JONES, A. G., ROSENQVIST, G., BERGLUND, A., ARNOLD, S. J., & AVISE, J. C. 2000. The Bateman gradient and the cause of sexual selection in a sex-role-reversed pipefish. *Proceedings of the Royal Society of London—Series B: Biological Sciences*, 267, 677–80.

JONES, B. C. & DUVAL, E. H. 2019. Mechanisms of social influence: a meta-analysis of the effects of social information on female mate choice decisions. *Frontiers in Ecology and Evolution*, 7, 390.

JORDAN, C. Y. & CONNALLON, T. 2014. Sexually antagonistic polymorphism in simultaneous hermaphrodites. *Evolution*, 68, 3555–69.

KAMIYA, T., O'DWYER, K., WESTERDAHL, H., SENIOR, A., & NAKAGAWA, S. 2014. A quantitative review of MHC-based mating preference: the role of diversity and dissimilarity. *Molecular Ecology*, 23, 5151–63.

KNELL, R. J. & WEBBERLEY, K. M. 2004. Sexually transmitted diseases of insects: distribution, evolution, ecology and host behaviour. *Biological Reviews*, 79, 557–81.

KOENE, J. M. 2017. Sex determination and gender expression: reproductive investment in snails. *Molecular Reproduction and Development*, 84, 132–43.

KOKKO, H., HEUBEL, K. U., & RANKIN, D. J. 2008. How populations persist when asexuality requires sex: the spatial dynamics of coping with sperm parasites. *Proceedings of the Royal Society of London—Series B: Biological Sciences*, 275, 817–25.

KOKKO, H. & JOHNSTONE, R. A. 2002. Why is mutual mate choice not the norm? Operational sex ratios, sex roles and the evolution of sexually dimorphic and monomorphic signalling. *Philosophical Transactions of the Royal Society B—Biological Sciences*, 357, 319–30.

KOMDEUR, J., OOREBEEK, M., VAN OVERVELD, T., & CUTHILL, I. C. 2005. Mutual ornamentation, age, and reproductive performance in the European starling. *Behavioral Ecology*, 16, 805–17.

KOTHE, E. 1996. Tetrapolar fungal mating types: sexes by the thousands. *FEMS Microbiology Reviews*, 18, 65–87.

KVARNEMO, C. & AHNESJÖ, I. 1996. The dynamics of operational sex ratios and competition for mates. *Trends in Ecology & Evolution*, 11, 404–8.

LACHLAN, R. F., RATMANN, O., & NOWICKI, S. 2018. Cultural conformity generates extremely stable traditions in bird song. *Nature Communications*, 9, 2417.

LALAND, K. N., ULLER, T., FELLMAN, M. W., STERELNY, K., MÜLLER, G. B., MOCZEK, A., JABLONKA, E., & ODLING-SMEE, J. 2015. The extended evolutionary synthesis: its structure, assumptions and predictions. *Proceedings of the Royal Society of London—Series B: Biological Sciences*, 282, 20151019.

LAMICHHANEY, S., HAN, F., WEBSTER, M. T., ANDERSSON, L., GRANT, B. R., & GRANT, P. R. 2018. Rapid hybrid speciation in Darwin's finches. *Science*, 359, 224–8.

LARSON, P. 2017. Brooding sea anemones (Cnidaria: Anthozoa: Actiniaria): paragons of diversity in mode, morphology, and maternity. *Invertebrate Biology*, 136, 92–112.

LEBAS, N. R., HOCKHAM, L. R., & RITCHIE, M. G. 2004. Sexual selection in the gift-giving dance fly, *Rhamphomyia sulcata*, favors small males carrying small gifts. *Evolution*, 58, 1763–72.

LEFEVRE, C. E. & SAXTON, T. K. 2017. Parental preferences for the facial traits of their offspring's partners can enhance parental inclusive fitness. *Evolution and Human Behavior*, 38, 546–51.

LEGETT, H. D., BARANOV, V. A., & BERNAL, X. E. 2018. Seasonal variation in abundance and diversity of eavesdropping frog-biting midges (Diptera, Corethrellidae) in a neotropical rainforest. *Ecological Entomology*, 43, 226–33.

LEHMANN, G. U. C. & LEHMANN, A. W. 2016. Material benefit of mating: the bushcricket spermatophylax as a fast uptake nuptial gift. *Animal Behaviour*, 112, 267–71.

LIKER, A., FRECKLETON, R. P., & SZEKELY, T. 2013. The evolution of sex roles in birds is related to adult sex ratio. *Nature Communications*, 4, 1–6.

LOSEY, G. S., Stanton, F. G., Telecky, T. M., Tyler III, W. A., & CLASS, Z. G. S. 1986. Copying others, an evolutionarily stable strategy for mate choice: a model. *American Naturalist*, 128, 653–64.

LOYAU, A., PETRIE, M., SAINT JALME, M., & SORCI, G. 2008. Do peahens not prefer peacocks with more elaborate trains? *Animal Behaviour*, 76, e5–9.

LÜPOLD, S., MANIER, M. K., PUNIAMOORTHY, N., SCHOFF, C., STARMER, W. T., LUEPOLD, S. H. B., BELOTE, J. M., & PITNICK, S. 2016. How sexual selection can drive the evolution of costly sperm ornamentation. *Nature*, 533, 535–38.

LÜSCHER, A. & WEDEKIND, C. 2002. Size-dependent discrimination of mating partners in the simultaneous hermaphroditic cestode *Schistocephalus solidus*. *Behavioral Ecology*, 13, 254–9.

LYON, B. E. & MONTGOMERIE, R. 2012. Sexual selection is a form of social selection. *Philosophical Transactions of the Royal Society B—Biological Sciences*, 367, 2266–73.

MAAN, M. E., HOFKER, K. D., VAN ALPHEN, J. J. M., & SEEHAUSEN, O. 2006. Sensory drive in cichlid speciation. *American Naturalist*, 167, 947–54.

MAKOWICZ, A. M. & SCHLUPP, I. 2013. The direct costs of living in a sexually harassing environment. *Animal Behaviour*, 85, 569–77.

MALLET, J. 2007. Hybrid speciation. *Nature*, 446, 279.

MANSER, A., KÖNIG, B., & LINDHOLM, A. 2015. Female house mice avoid fertilization by t haplotype incompatible males in a mate choice experiment. *Journal of Evolutionary Biology*, 28, 54–64.

MARLER, C. A., FORAN, C., & RYAN, M. J. 1997. The influence of experience on mating preferences of the gynogenetic Amazon molly. *Animal Behaviour*, 53, 1035–41.

MAXWELL, M. R., GALLEGO, K. M., & BARRY, K. L. 2010. Effects of female feeding regime in a sexually cannibalistic mantid: fecundity, cannibalism, and male response *in Stagmomantis limbata* (Mantodea). *Ecological Entomology*, 35, 775–87.

MAYNARD SMITH, J. & HARPER DAVID, D. 2003. *Animal Signals*. Oxford: Oxford University Press.

MAYR, E. 1940. Speciation phenomena in birds. *The American Naturalist*, 74, 249–78.

MAYR, E. 1963. *Animal Species and Evolution*. Cambridge, MA: Harvard University Press.

MCCOY, E., SYSKA, N., PLATH, M., SCHLUPP, I., & RIESCH, R. 2011. Mustached males in a tropical poeciliid fish: emerging female preference selects for a novel male trait. *Behavioral Ecology and Sociobiology*, 65, 1437–45.

MCGREGOR, P. K., OTTER, K., & PEAKE, T. M. 1999. Communication networks: receiver and signaller perspectives. *In*: ESPMARK, Y., AMUNDSEN, T., &

ROSENQVIST, G. (eds.) *Animal Signals: Signalling and Signal Design in Animal Communication*. Trondheim, Germany: Tapir Academic Press.

MENDELSON, T. C., GUMM, J. M., MARTIN, M. D., & CICCOTTO, P. J. 2018. Preference for conspecifics evolves earlier in males than females in a sexually dimorphic radiation of fishes. *Evolution*, 72, 337–47.

MENDELSON, T. C. & SHAW, K. L. 2012. The (mis)concept of species recognition. *Trends in Ecology & Evolution*, 27, 421–7.

MICHIELS, N. K. 1998. Mating conflicts and sperm competition in simultaneous hermaphrodites. *In*: BIRKHEAD, T. R. & MØLLER, A. P. (eds.) *Sperm Competition and Sexual Selection*. New York: Academic Press, pp. 219–53.

MILINSKI, M. 2006. The major histocompatibility complex, sexual selection, and mate choice. *Annual Review of Ecology Evolution and Systematics*, 37, 159–86.

MILINSKI, M. 2014. Arms races, ornaments and fragrant genes: the dilemma of mate choice in fishes. *Neuroscience and Biobehavioral Reviews*, 46, 567–72.

MOBLEY, K. B., KVARNEMO, C., AHNESJÖ, I., PARTRIDGE, C., BERGLUND, A., & JONES, A. G. 2011. The effect of maternal body size on embryo survivorship in the broods of pregnant male pipefish. *Behavioral Ecology and Sociobiology*, 65, 1169–77.

MØLLER, A. P. & BIRKHEAD, T. R. 1993. Certainty of paternity covaries with paternal care in birds. *Behavioral Ecology and Sociobiology*, 33, 261–8.

MØLLER, A. P. & CUERVO, J. J. 2000. The evolution of paternity and paternal care in birds. *Behavioral Ecology*, 11, 472–85.

MØLLER, A. P. & JENNIONS, M. D. 2001. How important are direct fitness benefits of sexual selection? *Naturwissenschaften*, 88, 401–15.

MORGADO-SANTOS, M., MAGALHÃES, M., VICENTE, L., & COLLARES-PEREIRA, M. 2018. Mate choice driven by genome in an allopolyploid fish complex. *Behavioral Ecology*, 29, 1359–70.

MORRIS, M. R. 1989. Female choice of large males in the treefrog *Hyla chrysoscelis*: the importance of identifying the scale of choice. *Behavioral Ecology and Sociobiology*, 25, 275–81.

NORDELL, S. E. & VALONE, T. J. 1998. Mate choice copying as public information. *Ecology Letters*, 1, 74–6.

NOSIL, P. 2012. *Ecological Speciation*. Oxford: Oxford University Press.

NOSIL, P., CRESPI, B. J., & SANDOVAL, C. P. 2002. Host-plant adaptation drives the parallel evolution of reproductive isolation. *Nature*, 417, 440–3.

NOURI, K. & BLUMSTEIN, D. T. 2019. Parasites are associated with noisy alarm calls. *Frontiers in Ecology and Evolution*, 7, 28.

NOZU, R., KOJIMA, Y., & NAKAMURA, M. 2009. Short term treatment with aromatase inhibitor induces sex change in the protogynous wrasse, *Halichoeres trimaculatus*. *General and Comparative Endocrinology*, 161, 360–4.

OLIVEIRA, R. F., LOPES, M., CARNEIRO, L. A., & CANARIO, A. V. M. 2001. Watching fights raises fish hormone levels. *Nature*, 409, 475–5.

OLIVEIRA, R. F., MCGREGOR, P. K., & LATRUFFE, C. 1998. Know thine enemy: fighting fish gather information from observing conspecific interactions. *Proceedings of the Royal Society of London—Series B: Biological Sciences*, 265, 1045–9.

ORFAO, I., BARBOSA, M., OJANGUREN, A. F., VICENTE, L., VARELA, S. A. M., & MAGURRAN, A. E. 2019. Me against who? Male guppies adjust mating behaviour according to their rival's presence and attractiveness. *Ethology*, 125, 399–408.

PARKER, G. A. 1974. Assessment strategy and the evolution of fighting behaviour. *Journal of Theoretical Biology*, 47, 223–43.

PARKER, G. A. 2006. Sexual conflict over mating and fertilization: an overview. *Philosophical Transactions of the Royal Society B—Biological Sciences*, 361, 235–59.

PARKER, G. A. 2016. The evolution of expenditure on testes. *Journal of Zoology*, 298, 3–19.

PARMA, P., VEYRUNES, F., & PAILHOUX, E. 2016. Sex reversal in non-human placental mammals. *Sexual Development*, 10, 326–44.

PARTRIDGE, C., AHNESJÖ, I., KVARNEMO, C., MOBLEY, K. B., BERGLUND, A., & JONES, A. G. 2009. The effect of perceived female parasite load on postcopulatory male choice in a sex-role-reversed pipefish. *Behavioral Ecology and Sociobiology*, 63, 345–54.

PEREZ-CAMACHO, L., GARCIA-SALGADO, G., REBOLLO, S., MARTINEZ-HESTERKAMP, S., & FERNANDEZ-PEREIRA, J. M. 2015. Higher reproductive success of small males and greater recruitment of large females may explain strong reversed sexual dimorphism (RSD) in the northern goshawk. *Oecologia*, 177, 379–87.

PETRIE, M. 1983. Mate choice in role-reversed species. In: BATESON, P. P. G. (ed.) *Mate Choice*. Cambridge, Cambridge University Press, pp. 167–79.

PETRIE, M. 1994. Improved growth and survival of offspring of peacocks with more elaborate trains. *Nature*, 371, 598–9.

PETRIE, M., HALLIDAY, T., & SANDERS, C. 1991. Peahens prefer peacocks with elaborate trains. *Animal Behaviour*, 41, 323–31.

PICHLER, P. 2007. Talking traditions of marriage—negotiating young British Bangladeshi femininities. *Womens Studies International Forum*, 30, 201–16.

PLATH, M., LIU, K., UMUTONI, D., GOMES-SILVA, G., WEI, J.-F., CYUBAHIRO, E., CHEN, B.-J., & SOMMER-

TREMBO, C. 2019. Predator-induced changes of male and female mating preferences: innate and learned components. *Current Zoology*, 65, 305–16

PLATH, M., MAKOWICZ, A. M., SCHLUPP, I., & TOBLER, M. 2007. Sexual harassment in live-bearing fishes (Poeciliidae): comparing courting and noncourting species. *Behavioral Ecology*, 18, 680–8.

PLATH, M., RICHTER, S., TIEDEMANN, R., & SCHLUPP, I. 2008. Male fish deceive competitors about mating preferences. *Current Biology*, 18, 1138–41.

PLATH, M. & SCHLUPP, I. 2008. Misleading mollies: the effect of an audience on the expression of mating preferences. *Communicative & Integrative Biology*, 1, 199–203.

PRESTON, B. T., STEVENSON, I. R., PEMBERTON, J. M., & WILSON, K. 2001. Dominant rams lose out by sperm depletion. *Nature*, 409, 681.

PROBST, F., FISCHBACHER, U., LOBMAIER, J. S., WIRTHMÜLLER, U., & KNOCH, D. 2017. Men's preferences for women's body odours are not associated with human leucocyte antigen. *Proceedings of the Royal Society of London—Series B: Biological Sciences*, 284, 20171830.

PROMEROVA, M., ALAVIOON, G., TUSSO, S., BURRI, R., & IMMLER, S. 2017. No evidence for MHC class II-based non-random mating at the gametic haplotype in Atlantic salmon. *Heredity*, 118, 563–7.

PUTS, D. A. 2010. Beauty and the beast: mechanisms of sexual selection in humans. *Evolution and Human Behavior*, 31, 157–75.

QIAO, Z., POWELL, J. E., & EVANS, D. M. 2018. MHC-dependent mate selection within 872 spousal pairs of European ancestry from the Health and Retirement Study. *Genes*, 9, 53.

QUINTANA, G. R., GUIZAR, A., RASSI, S., & PFAUS, J. G. 2018. First sexual experiences determine the development of conditioned ejaculatory preference in male rats. *Learning & Memory*, 25, 522–32.

QUIROZ, P. A. 2013. From finding the perfect love online to satellite dating and "loving-the-one-you're near": a look at Grindr, Skout, Plenty of Fish, Meet Moi, Zoosk and assisted serendipity. *Humanity & Society*, 37, 181–5.

ROBERTS, N. S. & MENDELSON, T. C. 2017. Male mate choice contributes to behavioural isolation in sexually dimorphic fish with traditional sex roles. *Animal Behaviour*, 130, 1–7.

RODD, F. H., HUGHES, K. A., GRETHER, G. F., & BARIL, C. T. 2002. A possible non-sexual origin of mate preference: are male guppies mimicking fruit? *Proceedings of the Royal Society of London—Series B: Biological Sciences*, 269, 475–81.

ROSENQVIST, G. & BERGLUND, A. 2011. Sexual signals and mating patterns in Syngnathidae. *Journal of Fish Biology*, 78, 1647–61.

ROSENTHAL, G. G. 2017. *Mate Choice*. Princeton, NJ: Princeton University Press.

ROSENTHAL, G. G. & RYAN, M. J. 2011. Conflicting preferences within females: sexual selection versus species recognition. *Biology Letters, 7,* 525–7.

ROSSELLÓ-MORA, R. & AMANN, R. 2001. The species concept for prokaryotes. *FEMS Microbiology Reviews, 25,* 39–67.

ROWE, L. 1994. The costs of mating and mate choice in water striders. *Animal Behaviour, 48,* 1049–56.

RUNDLE, H. D. & NOSIL, P. 2005. Ecological speciation. *Ecology Letters, 8,* 336–52.

RYAN, M. J., AKRE, K. L., & KIRKPATRICK, M. 2007. Mate choice. *Current Biology, 17,* 313.

RYAN, M. J. & CUMMINGS, M. E. 2013. Perceptual biases and mate choice. *In:* FUTUYMA, D. J. (ed.) *Annual Review of Ecology, Evolution, and Systematics, 44,* 437–59.

RYAN, M. J. & KEDDY-HECTOR, A. 1992. Directional patterns of female mate choice and the role of sensory biases. *American Naturalist,* 4–35.

RYAN, M. J., PAGE, R. A., HUNTER, K. L., & TAYLOR, R. C. 2019. "Crazy love": nonlinearity and irrationality in mate choice. *Animal Behaviour, 147,* 189–98.

RYAN, M. J. & RAND, A. S. 1993. Species recognition and sexual selection as a unitary problem in animal communication. *Evolution, 47,* 647–57.

SAKALUK, S. K., DUFFIELD, K. R., RAPKIN, J., SADD, B. M., & HUNT, J. 2019. The troublesome gift: the spermatophylax as a purveyor of sexual conflict and coercion in crickets. *Advances in the Study of Behavior, 51,* 1–30.

SANDVIK, M., ROSENQVIST, G., & BERGLUND, A. 2000. Male and female mate choice affects offspring quality in a sex-role-reversed pipefish. *Proceedings of the Royal Society of London—Series B: Biological Sciences, 267,* 2151–5.

SCHAMEL, D., TRACY, D. M., & LANK, D. B. 2004. Male mate choice, male availability and egg production as limitations on polyandry in the red-necked phalarope. *Animal Behaviour, 67,* 847–53.

SCHLUPP, I. 2005. The evolutionary ecology of gynogenesis. *Annual Review of Ecology Evolution and Systematics, 36,* 399–417.

SCHLUPP, I. 2009. Behavior of fishes in the sexual/unisexual mating system of the Amazon molly (*Poecilia formosa*). *In:* BROCKMANN, H. J., ROPER, T. J., NAGUIB, M., WYNNEEDWARDS, K. E., MITANI, J. C., & SIMMONS, L. W. (eds.) *Advances in the Study of Behavior, Vol. 39.* Amsterdam: Elsevier, pp. 153–83.

SCHLUPP, I. 2018. Male mate choice in livebearing fishes: an overview. *Current Zoology, 64,* 393–403.

SCHLUPP, I., MARLER, C., & RYAN, M. J. 1994. Benefit to male sailfin mollies of mating with heterospecific females. *Science, 263,* 373–4.

SCHLUPP, I., PARZEFALL, J., EPPLEN, J. T., NANDA, I., SCHMID, M., & SCHARTL, M. 1992. Pseudomale

behavior and spontaneous masculinization in the all-female teleost *Poecilia formosa* (Teleostei, Poeciliidae). *Behaviour, 122,* 88–104.

SCHLUPP, I., RIESCH, R., TOBLER, M., PLATH, M., PARZEFALL, J., & SCHARTL, M. 2010. A novel, sexually selected trait in poeciliid fishes: female preference for mustache-like, rostral filaments in male *Poecilia sphenops. Behavioral Ecology and Sociobiology, 64,* 1849–55.

SCHLUPP, I. & RYAN, M. J. 1997. Male sailfin mollies (*Poecilia latipinna*) copy the mate choice of other males. *Behavioral Ecology, 8,* 104–7.

SCHULTE, L. M. & LOTTERS, S. 2013. The power of the seasons: rainfall triggers parental care in poison frogs. *Evolutionary Ecology, 27,* 711–23.

SCHWAGMEYER, P. L., ST CLAIR, R. C., MOODIE, J. D., LAMEY, T. C., SCHNELL, G. D., & MOODIE, M. N. 1999. Species differences in male parental care in birds: a reexamination of correlates with paternity. *Auk, 116,* 487–503.

SEARCY, W. A. & NOWICKI, S. 2005. *The Evolution of Animal Communication: Reliability and Deception in Signaling Systems.* Princeton, NJ: Princeton University Press.

SEARCY, W. A. & NOWICKI, S. 2019. Birdsong learning, avian cognition and the evolution of language. *Animal Behaviour, 151,* 217–27.

SETCHELL, J. M., RICHARDS, S. A., ABBOTT, K. M., & KNAPPE, L. A. 2016. Mate-guarding by male mandrills (*Mandrillus sphinx*) is associated with female MHC genotype. *Behavioral Ecology, 27,* 1756–66.

SHARANGPANI, M. 2010. Browsing for bridegrooms: matchmaking and modernity in Mumbai. *Indian Journal of Gender Studies, 17,* 249–76.

SHELDON, B. C. 2002. Relating paternity to paternal care. *Philosophical Transactions of the Royal Society B—Biological Sciences, 357,* 341–50.

SHIFFERMAN, E. M. 2012. It's all in your head: the role of quantity estimation in sperm competition. *Proceedings of the Royal Society of London—Series B: Biological Sciences, 279,* 833–40.

SIMMONS, L. 1995. Courtship feeding in katydids (Orthoptera: Tettigoniidae): investment in offspring and in obtaining fertilizations. *The American Naturalist, 146,* 307–15.

SIN, Y. W., ANNAVI, G., NEWMAN, C., BUESCHING, C., BURKE, T., MACDONALD, D. W., & DUGDALE, H. L. 2015. MHC class II-assortative mate choice in European badgers (*Meles meles*). *Molecular Ecology, 24,* 3138–50.

SONERUD, G. A., STEEN, R., SELÅS, V., AANONSEN, O. M., AASEN, G. H., FAGERLAND, K. L., FOSSÅ, A., KRISTIANSEN, L., LØW, L. M., RØNNING, M. E., SKOUEN, S. K., ASAKSKOGEN, E., JOHANSEN, H. M., JOHNSEN, J. T., KARLSEN, L. I., NYHUS, G. C., RØED, L. T., SKAR, K., SVEEN, B. A., TVEITEN, R., &

SLAGSVOLD, T. 2014. Evolution of parental roles in provisioning birds: diet determines role asymmetry in raptors. *Behavioral Ecology*, 25, 762–72.

STREET, S. E., MORGAN, T. J., THORNTON, A., BROWN, G. R., LALAND, K. N., & CROSS, C. P. 2018. Human mate-choice copying is domain-general social learning. *Scientific Reports*, 8, 1–7.

STUTT, A. D. & SIVA-JOTHY, M. T. 2001. Traumatic insemination and sexual conflict in the bed bug *Cimex lectularius*. *Proceedings of the National Academy of Sciences*, 98, 5683–7.

SUMMERS, K. 1989. Sexual selection and intra-female competition in the green poison-dart frog, *Dendrobates auratus*. *Animal Behaviour*, 37, 797–805.

SUMMERS, K. & TUMULTY, J. 2014. Parental care, sexual selection, and mating systems in neotropical poison frogs. *In*: MACEDO, R. H. & MACHADO, G. (eds.) *Sexual Selection: Perspectives and Models from the Neotropics*. Amsterdam: Elsevier.

SUOMALAINEN, E., SAURA, A., & LOKKI, J. 1987. *Cytology and Evolution in Parthenogenesis*. Boca Raton, FL: CRC Press.

SWELLER, J. 2011. Cognitive load theory. *In*: MESTRE, J. P. & ROSS, B. H. (eds.) *Psychology of Learning and Motivation, Vol. 55*. Amsterdam: Elsevier.

SWIERK, L., TENNESSEN, J. B., & LANGKILDE, T. 2015. Sperm depletion may not limit male reproduction in a capital breeder. *Biological Journal of the Linnean Society*, 116, 684–90.

TAKAHASHI, M., ARITA, H., HIRAIWA-HASEGAWA, M., & HASEGAWA, T. 2008. Peahens do not prefer peacocks with more elaborate trains. *Animal Behaviour*, 75, 1209–19.

TALLAMY, D. W. 2000. Sexual selection and the evolution of exclusive paternal care in arthropods. *Animal Behaviour*, 60, 559–67.

TATARNIC, N. J., CASSIS, G., & SIVA-JOTHY, M. T. 2014. Traumatic insemination in terrestrial arthropods. *Annual Review of Entomology*, 59, 245–61.

TINBERGEN, N. 1936. The function of sexual fighting in birds; and the problem of the origin of "territory." *Bird-banding*, 7, 1–8.

TRIVERS, R. (ed.) 1972. *Parental Investment and Sexual Selection*. Chicago, IL: Aldine.

TUTTLE, E. M. 2003. Alternative reproductive strategies in the white-throated sparrow: behavioral and genetic evidence. *Behavioral Ecology*, 14, 425–32.

TUTTLE, E. M., BERGLAND, A. O., KORODY, M. L., BREWER, M. S., NEWHOUSE, D. J., MINX, P., STAGER, M., BETUEL, A., CHEVIRON, Z. A., & WARREN, W. C. 2016. Divergence and functional degradation of a sex chromosome-like supergene. *Current Biology*, 26, 344–50.

ULLER, T. & JOHANSSON, L. C. 2003. Human mate choice and the wedding ring effect—are married men more attractive? *Human Nature—an Interdisciplinary Biosocial Perspective*, 14, 267–76.

URSPRUNG, E., RINGLER, M., JEHLE, R., & HOEDL, W. 2011. Strong male/male competition allows for non-choosy females: high levels of polygynandry in a territorial frog with paternal care. *Molecular Ecology*, 20, 1759–71.

VAKIRTZIS, A. & ROBERTS, S. C. 2010. Mate quality bias: sex differences in humans. *Annales Zoologici Fennici*, 47, 149–57.

VAN DEN BERG, P., FAWCETT, T. W., BUUNK, A. P., & WEISSING, F. J. 2013. The evolution of parent-offspring conflict over mate choice. *Evolution and Human Behavior*, 34, 405–11.

VENNER, S., BERNSTEIN, C., DRAY, S., & BEL-VENNER, M. C. 2010. Make love not war: when should less competitive males choose low-quality but defendable females? *American Naturalist*, 175, 650–61.

VINCENT, A., AHNESJÖ, I., BERGLUND, A., & ROSENQVIST, G. 1992. Pipefishes and seahorses: are they all sex role reversed? *Trends in Ecology & Evolution*, 7, 237–41.

WADE, M. J. & SHUSTER, S. M. 2005. Don't throw Bateman out with the bathwater! *Integrative and Comparative Biology*, 45, 945–51.

WANG, D., FORSTMEIER, W., VALCU, M., DINGEMANSE, N., BULLA, M., BOTH, C., DUCKWORTH, R. A., KIERE, L. M., KARELL, P., & ALBRECHT, T. 2019. Scrutinizing assortative mating in birds. *PLoS Biology*, 17, e3000156.

WAYNFORTH, D. 2007. Mate choice copying in humans. *Human Nature—an Interdisciplinary Biosocial Perspective*, 18, 264–71.

WEATHERHEAD, P. J. & ROBERTSON, R. J. 1979. offspring quality and the polygyny threshold: "the sexy son hypothesis." *The American Naturalist*, 113, 201–8.

WEDEKIND, C., SEEBECK, T., BETTENS, F., & PAEPKE, A. J. 1995. MHC-dependent mate preferences in humans. *Proceedings of the Royal Society of London—Series B: Biological Sciences*, 260, 245–9.

WEEKS, S. C., BENVENUTO, C., & REED, S. K. 2006. When males and hermaphrodites coexist: a review of androdioecy in animals. *Integrative and Comparative Biology*, 46, 449–64.

WERREN, J. H., BALDO, L., & CLARK, M. E. 2008. Wolbachia: master manipulators of invertebrate biology. *Nature Reviews Microbiology*, 6, 741.

WHITFIELD, D. P. 1990. Male choice and sperm competition as constraints on polyandry in the red-necked phalarope *Phalaropus lobatus*. *Behavioral Ecology and Sociobiology*, 27, 247–54.

WHITTINGHAM, L. A., SHELDON, F. H., & EMLEN, S. T. 2000. Molecular phylogeny of jacanas and its implications for morphologic and biogeographic evolution. *Auk,* 117, 22–32.

WIDEMO, M. S. 2006. Male but not female pipefish copy mate choice. *Behavioral Ecology,* 17, 255–9.

WILLIAMS, S. R. & PIKE, I. L. 2002. What does love have to do with it? Perceived stress levels and health in love marriages vs. arranged marriages in a tribal population of India. *American Journal of Human Biology,* 14, 136–7.

WILSON, A. B., VINCENT, A., AHNESJÖ, I., & MEYER, A. 2001. Male pregnancy in seahorses and pipefishes (Family Syngnathidae): rapid diversification of paternal brood pouch morphology inferred from a molecular phylogeny. *Journal of Heredity,* 92, 159–66.

WING, S. R. 1988. Cost of mating for female insects: risk of predation in *Photinus collustrans* (Coleoptera: Lampyridae). *The American Naturalist,* 131, 139–42.

WITTE, K. & RYAN, M. J. 2002. Mate choice copying in the sailfin molly, *Poecilia latipinna*, in the wild. *Animal Behaviour,* 63, 943–9.

WONG, B. B. & JENNIONS, M. D. 2003. Costs influence male mate choice in a freshwater fish. *Proceedings of the Royal Society of London—Series B: Biological Sciences,* 270, S36–8.

WOODROFFE, R. & VINCENT, A. 1994. Mother's little helpers: patterns of male care in mammals. *Trends in Ecology & Evolution,* 9, 294–7.

WU, K. R., CHEN, C. S., MOYZIS, R. K., NUNO, M., YU, Z. X., & GREENBERGER, E. 2018. More than skin deep: major histocompatibility complex (MHC)-based attraction among Asian American speed-daters. *Evolution and Human Behavior,* 39, 447–56.

YAKUSHKO, O. & RAJAN, I. 2016. Global love for sale: divergence and convergence of human trafficking with "mail order brides" and international arranged marriage phenomena. *Women & Therapy,* 40, 190–206.

YOSHIZAWA, K., FERREIRA, R. L., LIENHARD, C., & KAMIMURA, Y. 2019. Why did a female penis evolve in a small group of cave insects? *BioEssays,* https://doi.org/10.1002/bies.201900005.

YOSHIZAWA, K., FERREIRA, R. L., YAO, I., LIENHARD, C., & KAMIMURA, Y. 2018a. Independent origins of female penis and its coevolution with male vagina in cave insects (Psocodea: Prionoglarididae). *Biology Letters,* 14, 20180533.

YOSHIZAWA, K., KAMIMURA, Y., LIENHARD, C., FERREIRA, R. L., & BLANKE, A. 2018b. A biological switching valve evolved in the female of a sex-role reversed cave insect to receive multiple seminal packages. *eLife,* 7, e39563.

CHAPTER 3

Examples of Male Mate Choice

3.1 Brief outline of the chapter

In this chapter, what interests me most is how often male mate choice has already been documented, independent of the underlying mechanism. I am not concerned with the origin of the description: some authors express some degree of surprise that they found male preferences; other studies are motivated by theory. I also want to highlight that there is a continuum from no male contribution to the offspring to male contributions that are larger than the female contribution. Furthermore, there are differences in female quality at different levels, which can contribute to the evolution of male choice. There are many studies that infer differences in female fecundity as underlying male choice, but females can differ in many more aspects—just like males.

3.2 Limitations of this chapter

In this chapter I want to mainly review known examples of male mate choice. I tried to find as many as possible. Most likely, my list is not complete, but it will hopefully at least indicate where more research is needed. I am deliberately just listing studies while providing limited details and background. A brief discussion of the main putative avenues for the evolution of male mate choice can be found in the following chapters. I am providing a list of examples (Table 3.1), which is not sorted taxonomically but according to the putative benefit to male mate choice according to the authors of the studies. The chapter, however, is organized according to taxon. Where possible, I also report studies

that searched for male mate choice but did not find any evidence for it. I also just list the studies I found with no claim to resolving the question of how and why male mate choice evolved in these cases.

3.3 Male mate choice viewed taxonomically

Research in general has a strong taxonomic bias, and sexual selection and mate choice are no different. There are many different reasons for these biases. One is that the scientific community often chooses so-called model systems to study the questions at hand. Much of the current biomedical research is done on rats and mice, rodents that are only somewhat similar to humans in many aspects, but they can easily be held and bred in large numbers in small cages. A fish model system in developmental biology, the zebrafish (*Danio rerio*) is probably a model organism because it has translucent eggs, a trait that is actually not common in fishes. Similarly, *Drosophila melanogaster* may be a model organism, not because they are typical for insects, but because they are easy to keep in the laboratory. Nowadays many resources are available for model systems such as *Drosophila*, making it even easier and ever more appealing to use a model system. However, just the notion that one or a few species can represent a whole group is problematic, especially for a group as speciose and diverse as insects. Once model organisms are established it becomes increasingly easy to use them because so much is already known about them and the community of peers with the same interest is large. In the field of animal behavior the bias is heavily in

Male Choice, Female Competition, and Female Ornaments in Sexual Selection. Ingo Schlupp, Oxford University Press (2021). © Ingo Schlupp (2021). DOI: 10.1093/oso/9780198818946.003.0003

favor of vertebrates (Rosenthal et al., 2017), presumably because formative early work in both ethology and behavioral ecology used vertebrates. Yet another potential problem is that we are focusing on hypothesis-driven research only, thereby running the risk of ignoring the natural history of the species we work with (Travis, 2020). There is also geographic bias, with more research done in temperate zones (Culumber et al., 2019). Clearly, there are several sociological, psychological, socioeconomic and political reasons for such biases, but it is important to understand how they may limit which questions we ask, how we ask them, and who asks them. Acknowledging these biases, however, helps us put our research into perspective and identify important gaps and problems. Exploring the existing literature according to taxonomic groups is made possible by databases like Web of Knowledge, or Google Scholar, and aided by the Tree of Life Project (Maddison and Schulz, 2007). These tools will retrieve records based on the keywords and search syntax used. This is of course also a potential weakness of such searches because poor keywords will return poor data. For example, if a paper on male mate choice does use keywords pointing to male mating preferences, I may have missed it. However, a taxonomic overview is more limited by the nature of what has been explored scientifically than the taxa that would be worth exploring. Hence, my review is completely opportunistic. Nonetheless, I think the examples listed here can be useful in documenting trends of what we already know and identify more clearly what we do not know. The species names referred to in the text are the ones used by the authors and may not reflect the latest taxonomy.

3.4 Plants

Sexual selection in higher plants is still a little enigmatic but has received considerable attention soon after the paradigm shift in biology that led to a gene selection view (Dawkins, 1976). In their book, Willson and Burley (1983) dubbed the field sociobotany, a word-play on the—then new—term sociobiology (Wilson, 1975). The basis for sexual selection to occur, that is, differential investment into gametes, is clearly present in plants as well; indeed, it

may be even more extreme in plants than in animals. Andersson suggested the lack of compensatory male investment as the reason for this (Andersson, 1994). Around the same time Arnold suggested a unified theory and terminology for sexual selection in plants and animals (Arnold, 1994b). Plants produce male gametes that are mobile, pollen, and female gametes that need to be found by male gametes, ovules. It also seems clear that the investment in ovules is much larger than in pollen. The ground for sexual selection to happen is laid and was already worked out by Bateman (1948). Yet, the effects differ markedly in animals and plants (Marshall and Folsom, 1991; Arnold, 1994a). In plants nonrandom mating is found, but the mechanisms are difficult to document (Dorken and Perry, 2017), partly at least because the fate of pollen is difficult to observe. This nonrandom mating is possibly the effect of some form of female choice, but this is far from clear. The mechanisms behind this potential female choice are also not clear (Marshall and Folsom, 1991). There seems to be agreement, however, that sexual selection is occurring in plants (Willson, 1994; Skogsmyr and Lankinen, 2002; Dorken and Perry, 2017). Nonetheless, it is perplexing that the effects differ so strongly in plants and animals. While female choice and/or male competition lead to ornamental traits in animals, such effects are more difficult to document in plants. Conspicuous and costly traits in pollinators are thought to be under sexual selection, but the evolution of these traits often happens via an intermediary, namely pollinating animals with interesting effects on the plants (Caruso et al., 2019). By producing traits that are more attractive to pollinators male plants may be able to increase their fitness (Skogsmyr and Lankinen, 2002). This may be most conspicuous in orchids that mimic female insects to attract insect males to "mate" with them, which leads to the pollination of the plant (Schiestl et al., 1999; Gaskett, 2011) (Figure 3.1). While there is some evidence for female choice (Betts et al., 2015) and male competition (Andersson, 1994), in particular in some milkweeds (Cocucci et al., 2014), there is little to no evidence for male choice or female competition (Burd, 1994). However, there is evidence for Bateman gradients in an annual,

Figure 3.1 *Andrena pilipes*, a bee species, attempting copulation with an *Ophrys garganica* orchid, thereby pollinating the plant (photo credit: Florian Schiestl).

dioecious herb, *Mercurialis annua*, being much steeper in males (Tonnabel et al., 2019). This suggests that sexual selection operates similarly to animals.

3.5 Fungi

Fungi are very different from plants and animals in many ways. Many fungi, like yeast, are single celled and sexuality involves same-sized cells, with no sex roles distinguishable. In multicellular fungi, such as filamentous ascomycetes or Agaricomycotina (mushroom-forming basidiomycetes), however, often mating types that resemble eggs and sperm can be distinguished. Consequently, there is potential for sexual selection to occur. Both female choice and male competition have been reported, but no male choice and female competition yet (Nieuwenhuis and Aanen, 2012). One has to keep in mind though, that fungi often have several mating types, and do not necessarily conform to the concept of two sexes.

3.6 Animals

3.6.1 Vertebrates

Within the animals, I arbitrarily decided to start listing vertebrate examples followed by invertebrate examples. This is partly because I think (hopefully correctly) that many of the examples stem from this

group and is not meant to diminish the importance of invertebrates.

Elasmobranchii

Not much is known about mate choice in sharks, and other elasmobranchids. One study looked for mate choice in the spot-tail shark, *Carcharhinus sorrah*, but no male mate choice for body size, parasite burden, or mean heterozygosity was found (Almojil, 2020).

Teleost fishes

Teleost fishes provide a number of examples for male mate choice. This is not really surprising as the diversity of sexual behaviors and mating systems in fishes is enormous. They are also the most specious group of all vertebrates (Betancur-R et al., 2013), with an estimate of well over 30,000 species. An early review of the literature on male mate choice in fishes was published by Sargent et al. (1986). The relatively rich literature on livebearing fishes (Poeciliidae) was reviewed in Schlupp (2018). In this taxonomic group there is no male investment in the offspring beyond the ejaculate, so likely male choice evolved because of differences in female quality. In a study designed to specifically test for both female and male choice (Basolo, 2004) in a livebearing fish from Costa Rica, *Brachyraphis rhapdophora*, Basolo found female choice but not male choice. Several studies on guppies (*Poecilia reticulata*) reported that males prefer larger females and adjust their preference to the presence of rivals (Dosen and Montgomerie, 2004a, 2004b). Godin and several of his students explored many aspects of male mate choice in guppies and other species (Herdman et al., 2004; Hoysak and Godin, 2007; Auld et al., 2015, 2017). In *Xiphophorus malinche*, another livebearing fish from Mexico, males preferred larger females. Interestingly, males of different size classes differed in their strength of preference. Also, in this mating system females of different sizes prefer different males: smaller females prefer symmetrical males, while larger females prefer asymmetrical males. These two types of males, however, did not differ in their preferences (Tudor and Morris, 2009). In a small species flock of pupfish (*Cyprinodon variegatus, Cyprinodon desquamator*, and *Cyprinodon brontotheroides*) from the Bahamas, both males and females preferred conspecifics, thereby maintaining species boundaries (West and Kodric-Brown, 2015).

There are some special cases: in some species females use sperm from males of a different species to trigger embryogenesis (Chapter 2). In males that provide sperm for unisexual females, such as the Amazon molly (*Poecilia formosa*) and *Poeciliopsis*, another genus in which sperm-dependent parthenogenesis evolved, male preferences for conspecific, sexual females are common (Keegan-Rogers, 1984; Schlupp, 2018b).

Male choice can be influenced by personalities and individual differences in a mosquitofish (*Gambusia affinis*) (Chen et al., 2018). A study using feral mosquitofish (*Gambusia holbrooki*) from Australia found differences in male preferences for larger females based on the mode of presentation (Head et al., 2015). Males showed stronger preferences in simultaneous encounters, which is also predicted by theory because males can afford to be less choosy in sequential encounters as there is no opportunity cost (Barry and Kokko, 2010). Furthermore, this seems to be a general phenomenon that needs more exploration (Dougherty and Shuker, 2014).

In another freshwater fish, *Percina roanoka*, males did not show a preference for larger females, but neither did females (Ciccotto et al., 2014). Interestingly, males of the pygmy halfbeak (*Dermogenys collettei*), from the family Hemiramphidae, prefer females with a larger orange ornament (the gravid spot), but not larger females *per se* (Ogden et al., 2019).

In darters, a group of colorful and diverse fishes from North America, male mate choice has been documented multiple times. In one species, the orangethroat darter (*Etheostoma spectabile*) males did not prefer larger females (Pyron, 1996). Pyron speculated that the reason for this might be a lack of differences in female quality. In *Etheostoma caeruleum*, the rainbow darter, however, a male preference for larger females was reported (Soudry et al., 2020). In other species, male choice is relatively strong and plays an important role in speciation (Martin and Mendelson, 2016; Moran et al., 2017).

In cichlids, a very species-rich and ecologically diverse family of fishes, males often invest into off-spring by providing paternal care (Chapter 5). This can take the form of males guarding and defending nests and broods, but also of mouthbrooding in some species, where males raise the young in their mouth cavity. However, male mate choice has been reported mostly from species without major male investment in offspring. For example, *Neochromis omnicaeruleus*, an African species from Lake Victoria, shows clear individual male preferences for female color morphs (Pierotti et al., 2009). This is interpreted in the context of incipient speciation: if males consistently and heritably prefer certain females, this may lead to new species. A preference for larger females was reported for another cichlid from east Africa, the convict cichlid (*Cichlasoma nigrofasciatum*) (Nuttall and Keenleyside, 1993), in this case leading to assortative mating (Beeching and Hopp, 1999). Interestingly, male preferences may be limited by female aggression (Bloch et al., 2016). In another cichlid, *Astatotilapia flaviijosephi*, a male preference for larger females was found both when the females were presented simultaneously (Werner and Lotem, 2003) and sequentially (Werner and Lotem, 2006), providing congruent evidence for male mate choice independent of methodology. In other examples, the way the stimuli were presented led to differences in the results. This highlights the importance of methodology for placing results into context.

Male and female mate choice has been well explored in another cichlid from Africa, the small, cavity-nesting *Pelvicachromis taeniatus* (slender krib). In this species, females are very colorful, and their purple pelvic fins are classified as an ornament, which correlates with maturity and fecundity (Baldauf et al., 2011) (see Chapter 7). Both sexes show choice (Scherer et al., 2018), which in this system is known to be quite complex: both males and females prefer larger partners (Baldauf et al., 2010), but also relatives (interestingly, they actively inbreed). These two preferences can be in conflict (Thünken et al., 2012), and the two sexes resolve this differently. Reflecting an unusual tradeoff, males value size more than females but females prefer closer relatives (Thünken et al., 2012). Kin recognition is based on olfactory information (Thünken et al., 2014). Males also choose females based on a female ornament, the color and size of the female pelvic fins (Baldauf et al., 2011, 2010). However, a personality trait, boldness, was not important (Scherer and Schuett, 2018). Furthermore, females compete for males (Baldauf et al., 2011). Interestingly, females lose color when nesting, which might be correlated with a reduction of testosterone akin to what has been suggested in birds in researching the challenge hypothesis (Wingfield, 2017). This hypothesis posits that males should downregulate their testosterone levels while providing paternal care to mitigate potentially negative effects of high testosterone such as aggressiveness against the mate (Wingfield, 2017; Goymann et al., 2019). Males of another cichlid, the St. Peter's fish from Lake Kinaret (*Sarotherodon galilaeus*), also show preferences for larger females, just like females prefer larger males (Balshine-Earn, 1996).

In males of a popular pet species, the Siamese fighting fish (*Betta splendens*), a preference for conspecifics was found, and this preference is stronger than that of females (Justus and Mendelson, 2018). Another species that shows a male preference for females with more eggs is the Japanese medaka (*Oryzias latipes*) (Grant et al., 1995). Because we have many genomic resources for this species, more research would be much warranted, maybe discovering genes associated with male and female choice.

In a cyprinid fish, *Puntius titteya*, male preference for female coloration was found (Mieno and Karino, 2019). Males prefer females that are redder than other females and redness can be interpreted as a potential indicator for female quality. In the same species, females also prefer redder males (Fukuda and Karino, 2014).

In three-spined sticklebacks (*Gasterosteus aculeatus*), Candolin and Salesto (2009) found that males prefer larger females and court them more vigorously, but this preference can break down under male competition and also for low-quality males. Earlier studies came to the same conclusion and found a clear preference for larger females (Rowland, 1982; Rowland and Sevenster, 1985; Rowland, 1989; Kraak and Bakker, 1998), with a link between female size and fecundity (Sargent et al., 1986). Males also choose females based on information related to "head up posture," an indication that females are ready to spawn (Rowland and Sevenster,

1985; Bakker and Rowland, 1995). Ornamentation is also relevant in male choice (Rowland et al., 1991; Bakker and Rowland, 1995; Yong et al., 2018), but in a population from Canada, no male preference for female throat color was detected (Wright et al., 2015). In a different species, the brook stickleback (*Culaea inconstans*), males also prefer females showing nuptial coloration (McLennan, 1995). Male and female *G. aculeatus* also show differences in cognition. When confronted with spatial tasks, males and females respond differently (Keagy et al., 2019), but more work on this is needed. Males also differentiate—using olfaction—between females with and without experience with predators (Dellinger et al., 2018), showing fewer courtship displays toward predator-exposed and stressed females. This will be covered in more detail in Chapter 8, which discusses female ornamentation. Clearly in this system both partners choose (Kraak and Bakker, 1998).

Male Coho salmon (*Oncorhynchus kisutch*) prefer larger females, but their preference is influenced by their own size, leading to assortative mating. Furthermore, males prefer females that show more red body coloration (Foote, 1988; Foote and Larkin, 1988; Foote et al., 2004). Coho salmon show an interesting pattern of variability: there was strong evidence for a male preference for larger females but only in 1 of 2 years this was tested (Sargent et al., 1986). Male mate choice for larger females was also found in a small fish from Australia, *Pseudomugil signifier* (Wong and Jennions, 2003).

Additionally, in annual fishes, male mate choice for larger females has been documented. In *Austrolebias reicherti*, a small fish from South America, a visual male preference for larger females and a correlation between fecundity and size was found (Passos et al., 2019). Previous work had already shown female preferences for larger males and a relationship with operational sex ratio (OSR) where early in the season with an even sex ratio, females showed a preference for larger males, whereas late in the season as male numbers dwindle, this preference is not detectable anymore (Passos et al., 2014). Weakly electric fishes from Africa are morphologically very similar, but show different electrical signals (EOD), which males from at least one species, *Campylomormyrus compressirostris*, use in species recognition (Nagel et al., 2018).

Several marine fishes have also been studied. Gobies have been well studied in terms of male mate choice (Amundsen, 2018; Heubel, 2018). In one species, *Gobiusculus flavescens*, a small fish, a relatively weak preference for larger females was reported, which was attributed to limited variation in female size (Pelabon et al., 2003). In other words, if females are providing a homogeneous fitness return for males, selection on male choosiness is absent or weak. In the mangrove killifish, *Kryptolebias marmoratus*, with a very complex mating system, male choice was also found (Ellison et al., 2013). This is one of the few hermaphroditic vertebrates and the only known one to be capable of selfing. In addition to hermaphrodites, males are found in this species. While hermaphrodites are not choosy, males are and base their preferences on the major histocompatibility complex (MHC) (Ellison et al., 2013).

A study looking at both male and female preferences in marine redlip blennies (*Ophioblennius atlanticus*) found that both sexes prefer larger mating partners (Côté and Hunte, 1989). They also prefer older males (Côté and Hunte, 1993), although it is not clear what the potential adaptive benefit of this is. Preferences for older males have also been reported from other taxa, such as sparrows (*Passer domesticus*) and humans (*Homo sapiens*) (Buss and Schmitt, 2019). In another species of blenny, *Salaria pavo*, males are choosy in a population where breeding sites are rare (Almada et al., 1994, 1995). In this case OSR seems to be driving the dynamic. When breeding males are rare, they show choosiness and the females have to compete (Chapter 5).

Males of the ocellated wrasse (*Symphodus ocellatus*) sometimes refuse to mate with some females when too many sneaker males are present (Alonzo and Warner, 1999). This is male choice by rejection.

One species of wrasse presents an interesting situation of polymorphism. Males of a reef dwelling wrasse, *Thalassoma bifasciatum*, show two discrete spawning behaviors, group spawning and pair spawning. Males appear to express preference for larger females in group spawning, but not in pair spawning (Van den Berghe and Warner, 1989). This might point to a role for OSR in this context, although the preference of larger males for larger

females is also explained by fecundity. This species shows sex change and it would be interesting to know if preferences expressed as one sex are retained when an individual changes its sex.

In a damselfish species (*Stegastes leucosticus*) males also prefer larger females (Itzkowitz et al., 1998). In this study, however, males that mated multiple times with smaller females produced an equal number of offspring, making it unclear why the males have a preference for larger females. One could speculate that the preference for larger females might create a sequence of matings with the most beneficial females receiving the first matings.

Amphibians

Male parental care is widespread in frogs. It is not surprising therefore, that male mate choice has been documented several times in amphibians. Interestingly, it is not only found in species with male parental care, but also in species where neither females nor males provide parental care, echoing some of the literature on fishes where male choice is found without male investment (Schlupp, 2018). In a large comparative study (Vági et al., 2019) many transitions between male and female parental care were found, suggesting that sex roles are actually more fluid than usually considered and likely driven by ecological factors. The only parental care trait that was only found in females—for obvious reasons—was providing eggs as food for tadpoles, a trait that evolved several times independently (Fischer et al., 2019). Actually, given how often male parental care—and hence male investment—is found in anurans, there are relatively few accounts of male mate preferences. An example of male preferences in a species with no male investment is the wood frog, *Rana sylvatica*, where males prefer mating with larger females, although this did not lead to assortative mating (Berven, 1981). Also studying the American wood frog (renamed *Lithobates sylvaticus*), Swierk and Langkilde (2019) reported that males preferred larger females, but that offspring with preferred females were less fit than offspring from nonpreferred females (Swierk and Langkilde, 2019). This might be an example of males being able to show choice in a system where males invest very little, but not always succeeding in making adaptive choices. In a European species, *Rana dalmatina*,

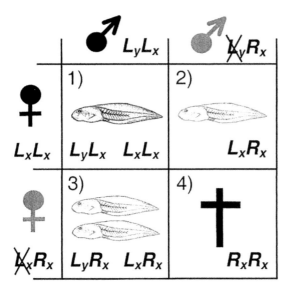

Figure 3.2 Possible mating combinations and resulting offspring (cells 1–4) in mixed populations of Rana lessonae (genotype LL) and R. esculenta (genotype LR). Note the lethal combination.

with an explosive breeding mating system (Wells, 1977), male mate choice was not detected, likely due to the intense scramble competition males are facing (Hettyey et al., 2005). In a species complex of European frog (*Rana esculenta* complex; now *Pelophylax esculentus*), which is characterized by a very complex, hemiclonal mating system, female choice was detected, but no male choice, although one of the potential genetic combinations is lethal (Engeler and Reyer, 2001) (Figure 3.2).

Male mate choice in some toads (genus *Bufo*) has been modeled by Krupa and colleagues (Krupa, 1995). Furthermore, in a subtropical *Bufo* species from China, *Bufo andrewsi*, males also preferred larger females (Liao and Lu, 2009). There also seemed to be an effect of OSR on choosiness.

Male parental care is the hallmark of midwife toads in the European genus *Alytes*. In *Alytes* species, males can carry eggs from several females, strung around their legs and protected by a strong toxin. Ecologically, it seems possible that this breeding system evolved owing to a lack of waterbodies for laying eggs. This is plausible because midwife toads are usually associated with dry habitats. Females compete for males in *Alytes obstetricans* (Verrell and Brown, 1993) and also in the Majorcan

midwife toad, *Alytes muletensis* (Bush and Bell, 1997). Curiously, females also produce calls (Dyson et al., 1998), and in *Alytes cisternasii*, females approach males to initiate mating (Márquez and Verrell, 1991). Males of *A. muletensis* show preferences for intermediate female calls (Bush et al., 1996). As is common in anurans, females prefer males with deeper voices (Márquez, 1995). Males with a lower fundamental frequency are larger, but this preference does not always lead to actual matings with larger males (Lea et al., 2003). Interestingly, males are reported to not always be willing to mate (Lea et al., 2003). This could, of course, be again male choice by rejection.

In the neotropical poison frogs in the families Dendrobatidae and Aromobatidae a majority of species show some kind of paternal care, which is otherwise not common in anurans (Summers and Tumulty, 2014). Although mainly famous for their toxicity and bright coloration, Dendrobatids also have an intriguing breeding behavior. Often, they lay small clutches of eggs, then guard or even feed them (Dugas et al., 2016), and subsequently transport the larvae on their backs to a stream upon hatching. Commonly, this transport is done by males (Summers and Tumulty, 2014). A phylogeny revealed that terrestrial breeding and male care are ancestral to the clade (Summers and McKeon, 2004). In the green poison frog (*Dendrobates auratus*) females show mate guarding of males they mated with, likely reflecting sexual conflict: males that mate multiply do so at a cost to female fitness, which the females try to thwart by aggressively guarding the male (Summers, 2014). Interestingly, many species are monomorphic, showing very similar ornamentation in males and females. Of course, this is likely warning coloration. Despite this it seems that a majority of the literature suggests mainly female choice as the mechanism for the evolution of ornaments in dendrobatids (Dugas et al., 2016). So far, male mate choice has not been reported for this group.

In an amazing case of what looks like parallel evolution, a tropical frog from Borneo, the smooth guardian frog (*Limnonectes palavanensis*) shows similar behavior: males care for the eggs and transport larvae to the water on their back (Vallejos et al., 2017). In this species, females court males by calling.

Just like males in many other species, the females form leks near males, which remain silent. After laying eggs, the females leave the males to care for the eggs. Although seemingly a good candidate for the evolution of male choice, there are no data for this, or for female–female competition for males.

In most newt species, males make heavy investments in their spermatophores. These are deposited—often after elaborate courtship—in the path of a female, which may or may not elect to pick up the spermatophore. Males may only be able to produce a certain number of spermatophores, providing a basis for the evolution of male mate choice. In the small salamander *Desmognathus santeelah*, males showed a preference for larger females when encountered simultaneously, but not when they met sequentially (Verrell, 1995). A similar result was obtained for *Desmognathus ochrophaeus* (Verrell, 1989) and *Desmognathus fuscus* (Verrell, 1994). Yet again, this raises interesting questions about the way methodology may influence behavioral outcomes, and interpretation of choice experiments. A study on another species of newt, *Notophthalmus viridescens*, also found a preference for larger females using a Y-maze. In this case the relevant information was either visual and/or olfactory (Verrell, 1985). The same author also reported male mate choice in the European smooth newt *Triturus vulgaris* (Verrell, 1986). A different species from the same genus (now renamed *Ichthyosaura*), *I. alpestris* was investigated for male choice for the red belly that is characteristic for females in this species. Males prefer redder females, but also females that were more responsive, even if less red. This seems to be a case where males trade off traits in females, redness and activity to maximize their reproductive success (Lüdtke and Foerster, 2018).

Reptiles

Lizards

Male mate choice has been reported in a number of reptile species. In European *Lacerta agilis*, male preference for female size was documented in a laboratory experiment (Olsson, 1993). Lindsay studied male mate choice, female competition, and ornamentation in an island population and found that males showed

clear signs of mating preferences and increased female aggression (W. Lindsay, pers. comm., March 22, 2019). In this species male copulation is short, but males are limited in copulation opportunities because they guard females for a long time.

In collared lizards (*Crotophytus collaris*) from Oklahoma, males (Figure 3.3) respond selectively to female ornaments (Baird, 2004) by courting ornamented females more.

In *Platysaurus broadleyi*, male preferences for larger females were detected (Whiting and Bateman, 1999; Wymann and Whiting, 2003). Interestingly—as in some other systems—males preferred larger females of an allopatric, but closely related species (*Platysaurus capensis*) over conspecific females. This demonstrated an important interaction between information used in species recognition and mate preference. Apparently, size is more important here than species identity (Wymann and Whiting, 2003).

Male mate choice in fence lizards, *Sceloporus undulatus*, uses multiple cues in mate preferences including body condition, and behavioral traits (Swierk et al., 2013). It is noteworthy that in the same system, females also show mate choice (Swierk et al., 2012). In a different species of fence lizard, *Sceloporus virgatus*, males show preference for a color ornament in females, but not for female size or reproductive state (Weiss and Dubin, 2018).

The widespread green anole, *Anolis carolinensis*, shows a preference for unfamiliar (and likely unmated) females (Orrell and Jenssen, 2002). In the brown anole (*Anolis sagrei*), a lizard from Cuba and

Figure 3.3 Collared lizard (*Crotophytus collaris*) male (photo credit: Ingo Schlupp).

the Bahamas (and elsewhere invasive), males were found to prefer familiar females, but not larger ones (Tokarz, 1992).

Female ornaments and male responses to them were studied in the Australian dragon lizard *Ctenophorus maculosus*. Males respond to the color ornaments of females but do not base their preference exclusively on them (Chan et al., 2009).

Overall, it seems interesting that in lizards female ornaments have been studied intensively and play an important role in male mate choice. There could be some taxonomic bias underlying this pattern.

Snakes

Male garter snakes (*Thamnophis sirtalis*) show a preference for larger females. In this system the preference is based on skin lipids and in an experiment, males courted larger females (Shine et al., 2006) and also skin extracts from larger females more often (LeMaster and Mason, 2002). Male mate choice was also suggested as a species-isolating mechanism in some species of *Thamnophis* (Shine et al., 2004).

Dinosaurs

Although somewhat speculative, and not based on direct observation, sexual selection (Knell et al., 2013) and male parental care has been suggested to exist in dinosaurs (Varricchio et al., 2008), dating the evolution of this behavior well before the evolution of modern birds. If male parental care is a condition that favors the evolution of male mate choice, that too might have been present in dinosaurs. Interestingly there is discussion of mutual sexual selection as the origin of ornaments in females of dinosaurs and other ancestors of birds (Hone et al., 2012), but of course it is difficult to come to firm conclusions based on the fossil record.

Birds

Birds are of great relevance here because many species show strong male investment, female ornamentation (including female song), and signs of both males and females choosing their partners (Komdeur et al., 2005). One would generally predict the evolution of male mate choice in this group.

Zebra finches (*Taeniopygia guttata*) are widely used in animal behavior studies. Some work found

that both males and females exercise mate choice (Wynn and Price, 1993). In another set of studies zebra finch males showed a preference for more fecund females after those had received a dietary supplement (Monaghan et al., 1996; Jones and Avise, 2001). This result, however, could not be repeated by others (Wang et al., 2017a, 2017b).

Some studies looked for effects of artificial ornaments on male preferences. For example, a male preference for a female with a new ornament was influenced by the social environment (Kniel et al., 2016). When exposed to novel and artificial ornaments, females showed signs of imprinting, but males did not (Witte and Sawka, 2003; Caspers and Witte, 2006). In another bird, the Javanese manakin (*Lonchura leucogastroides*), males rejected females that showed an artificial, new ornament (Witte and Curio, 1999). Furthermore, male Bengalese finches (*Lonchura striata domestica*) show preferences for individual females and adjust their courtship effort accordingly (Heinig et al., 2014).

In great tits (*Parus major*) from the Czech Republic, females with more pronounced plumage ornamentation, especially a larger black breast stripe and a white cheek, have heavier fledglings that survive better (Remeš and Matysioková, 2013). The authors suggest that this might be the basis for male mate choice.

Male dark eyed juncos, *Junco hyemalis*, distinguish between migrant and resident females during seasonal sympatry (Kimmitt et al., 2018). This seems to be adaptive through higher male reproductive success. One important effect is that male preferences are important in maintaining population differences, likely avoiding the cost of less fit hybrids. This is also found in other species, where males prefer a certain morph leading to assortative or disassortative mating and playing a role fairly similar to species recognition. Maybe this is the ancestral state for male choice: it is adaptive in the context of species and population recognition initially and then evolved into male preferences for other traits. In the polymorphic Gouldian finch, for example (*Erythrura gouldiae*), both sexes choose. Males show preferences for their own morph (Pryke and Griffith, 2007).

In a series of studies Amundsen and colleagues investigated the role of female ornaments in bluethroats (*Luscinia svecica*) from Europe and found that males prefer more ornamented females (Amundsen et al., 1997). Interestingly, females also preferred males with more symmetrical leg bands, which are widely used to mark individual birds (Hansen et al., 1999). A similar preference was also detected in females (Fiske and Amundsen, 1997). This is an interesting result showing how human activities can unwittingly influence mate choice and evolution.

House sparrows, *Passer domesticus*, also show signs of male mate choice (Veiga, 1990) via preferentially investing in the offspring of some females. There is also an interesting difference between older and younger males in terms of investment in ejaculates. Older males deliver three times more sperm as compared with younger males (Girndt et al., 2019). In house finches (*Carpodacus mexicanus*) males also prefer older females, but plumage coloration may play a role (Hill, 1993).

In a study looking at female mate choice, male mate choice and also male–male competition, and female–female competition, Johnson investigated pinyon jays (*Gymnorhinus cyanocephalus*) and found strong female choice for male coloration and testes size (Johnson, 1988a). Males, by contrast, preferred females that were heavier (Johnson, 1988b). She also noticed female competition for resources, while males competed for females, despite a generally monogamous mating system. Interestingly—and similar to the observations in house sparrows—males may exercise choice by simply ignoring females that want to mate with them (K. Johnson, pers. comm., February 21, 2019). Finally, in crested auklets, *Aethia cristatella*, a monomorphic, but ornamented Nordic bird, both sexes prefer mating partners that are more ornamented (Jones and Hunter, 1993) and interestingly the preferences in females were stronger than in males. The ornaments were manipulated experimentally and the experiment was conducted using dummies.

Mammals

Male mate choice has been reported for a number of mammals (Clutton-Brock, 2007, 2016), but for most species this appears not to have been investigated to date. Male parental care is relatively rare in mammals and consequently—assuming that this is a

driving force—male mate choice would not evolve often.

A study on domestic merino sheep (*Ovis aries*), revealed that rams allocated ejaculates unequally and preferred certain females, but showed little sign of sperm depletion (Synnott et al., 1981). This seems to be cryptic male choice. In feral Soay sheep (*Ovis aries*) from St. Kilda larger males preferred larger females, contributing to assortative mating (Preston et al., 2005). One interesting consequence of this assortative mating is that lighter females end up with more matings by inexperienced, lighter rams. Something similar has been reported for mountain goats, *Oreamnos americanus*, in which males seem to prefer older females (Mainguy and Cote, 2008; Mainguy et al., 2008). The underlying observations are from a field study, and association patterns may be caused by multiple factors, with male mate choice being just one of them. A study investigating mate copying in a species of gazelle from Saudi Arabia, *Gazella marica*, found male mate choice and short-duration male mate copying (Wronski et al., 2012).

In a study using bison (*Bison bison*), males were found to evaluate females based on reproductive potential and females of higher potential were preferred (Berger, 1989) (Figure 3.4).

In laboratory mice, males were found to actively avoid females infected with parasites (Gourbal and Gabrion, 2004). This is one of the few examples that investigated male choice in a context other than fecundity. Also, male mate choice based on olfaction has been reported as a mechanism to avoid

Figure 3.4 Bison (*Bison bison*) (photo credit: Ingo Schlupp).

hybridization between two subspecies of the house mouse, *Mus musculus musculus* and *Mus musculus domesticus* (Ganem et al., 2008). Not much is known about mate choice, let alone male mate choice, in the extinct Rhinogradentia (Stümpke, 1975).

A study using horses (*Equus ferus caballus*) found that males adjust the composition of their ejaculate relative to the reproductive status of the teaser mares they had smelled just before ejaculation (Jeannerat et al., 2017) and based on MHC (Jeannerat et al., 2018). This is another example of cryptic male choice.

Some studies exist from carnivores. The situation could be especially interesting in cooperative breeders, where only a small number of individuals reproduce. For example, in the banded mongoose (*Mungus mungo*), male mate choice was documented, likely driven by the fact that few mating partners are available at any given time and that females differ in reproductive output (Nichols et al., 2010). This highlights again the role of the OSR in male mate choice. Male mate choice is expressed as selective mate guarding. Also, in this species, males were found to prefer less related females, apparently to avoid inbreeding depression (Sanderson et al., 2015). This indicates that factors other than fecundity can be important in male mate choice.

In wolves (*Canis lupus*), both males and females show mating preferences (Rabb et al., 1967). Here, social rank is an important driver of the mating system and both sexes prefer higher-ranking mating partners.

In invasive racoons (*Procyon lotor*) in Germany, male and female preferences for dissimilar MHC were reported (Santos et al., 2017). This is one of the relatively few examples for male mate choice not credited to fecundity.

Spotted hyenas (*Crocuta crocuta*) live in complex social formations characterized by matrilines (Holekamp et al., 2012). Males prefer females of high rank, but also females close to the time of conception (Szykman et al., 2007). High-ranking males are known to consort with preferred females, helping them to defend against other males (East and Hofer, 2001).

Marine mammals are generally more difficult to study than other mammals, but male mate preferences were suggested in humpback whales (*Megaptera novaeangliae*) based on association data (Pack et al., 2012; Orbach et al., 2014). This preference is likely leading to assortative mating and larger calves, which have a higher probability of surviving (Craig et al., 2002).

In an early study of male mate choice, Schwagmeyer and Parker (1990) found evidence for male preferences in thirteen-lined ground squirrels (*Spermophilus tridecemlineatus*) based on costly male sperm production (Schwagmeyer and Parker, 1990) and avoidance of sperm competition. In this case males avoid females that have already been mated, presumably to avoid sperm competition.

Male mate choice is relatively well documented in primates. Kappeler provided a general overview of female and male mate choice, and also covered female competition (Kappeler, 2012). He argued that male mate choice is adaptive owing to the cost associated with production of sperm and searching for receptive females. Generally speaking, it seems that males prefer females that have higher fecundity, which often is associated with female rank (Keddy-Hector, 1992). However, in mandrills (*Mandrillus sphinx*) males selectively guarded females with a dissimilar MHC genotype (Setchell et al., 2016) and generally prefer high-ranking females. Exercising preferences through mate guarding seems to be a common mechanism in mammals (Setchell and Jean Wickings, 2006).

In a free-ranging population of ring-tailed lemurs (*Lemur catta*) from St. Catherines Island in Georgia, USA, males preferred older females with high fecundity and females from an age cohort (4–9 years), which are also associated with high fecundity and infant survival (Parga, 2006). This mating system is otherwise characterized by female choice and strong male–male competition, which manifests itself with adaptations and counter-adaptations like mating plugs (Parga, 2003). However, males invest strongly in mate guarding (Parga, 2010).

Chimpanzees (*Pan troglodytes*) also show signs of male mate choice. Males have a preference for older females, but the adaptive benefit of that is not clear yet (Muller et al., 2006). This is also relevant because it contrasts with the well-documented preference in human men for younger women, which consequently might be a derived trait in humans.

In humans, *Homo sapiens*, there is ample evidence for male mate choice. It seems likely that both paternal investment in children and variability in reproductive potential (or fecundity) play a role in its evolution. Many men invest heavily in their children, which should allow choosiness to evolve in humans. Yet, mothers seem to invest much more than most fathers based on the duration of pregnancy and relatively long childcare, which is often done predominantly by mothers. The literature on human behavior comes from a variety of fields, including humanities and social sciences, alongside with evolutionary psychology (Buss, 2015). This provides us with an especially rich understanding of the evolution and consequences of human behavior but has also been a source of controversy (Confer et al., 2010), and prolonged discussions of the roles of nature and nurture. Yet, humans are a great model organism for looking simultaneously at female and male choice, and male–male and female–female competition. In humans we have excellent information on all of these elements of sexual selection. Furthermore, for no other organism do we know this much about the role and consequences of social behavior in mate choice, and we are on the cusp of understanding the additional role of epigenetics in human sexual selection and evolution. In addition, we know that mating preferences translate into actual mating decisions (Conroy-Beam and Buss, 2016), which is not the case for all other animal models.

From what we know it is clear that women and men emphasize different traits in mate choice (Conroy-Beam et al., 2015). Men seem to prefer a set of traits in their partners that emphasize female reproductive potential. This includes indicators of youth, which strongly correlate with the number of future children, roughly the equivalent of fecundity in many studies of other animals. This has led—for example—to preferences for clear skin, and a good muscle tone in women. These preferences make sense from an evolutionary point of view (Conroy-Beam et al., 2015), but we also know that many of the preferences are additionally shaped by culture, and societal fads and fashions. One only has to compare the prevailing fashion ideals in Europe over the past decades or centuries to see that they seem to change randomly. However, as much as cultures may differ and quickly change, many pref-erences are consistent across cultures, suggesting an adaptive evolutionary origin of these preferences. This does not necessarily imply that these traits are currently still adaptive, some of which may have changed into evolutionary traps (Schlaepfer et al., 2002), but they could be.

Women prefer partners that are good providers for their children. This, too, can be modified by societal conditions. Much of modern sexual selection theory has been tested in humans, and generally there is confirmation of theory and concordance with observations in other animals, but mate choice in humans is exquisitely complex (Buss, 2015).

Another important feature of mate choice in humans is that it tends to lead to assortative mating. This is the case for physical traits, but also for traits like the education of prospective partners, highlighting the role of schools as a venue where people meet potential future partners (McClendon, 2018).

3.6.2 Invertebrates

Acanthocephala

Parasitic worms may not be the first place to look for male mate choice, but in a study by Lawlor and colleagues, a mating pattern in *Moniliformis moniliformis* was detected that was most consistent with male mate choice for more fecund females (Lawlor et al., 1990). These unpleasant animals infect a number of species, including rats, humans, and cockroaches, sometimes altering the behavior of the hosts (Gotelli and Moore, 1992; Poulin, 1994).

Rotatoria

Rotifers are another taxon for which only few examples of male mate choice are described. In *Brachionus plicatilis* males prefer younger females (Gómez and Serra, 1996), a pattern that might be widespread.

Mollusca

Many snails are simultaneous hermaphrodites (see Chapter 2) making them particularly great subjects to study mate choice (Leonard, 2006; Anthes et al., 2014), especially sperm trading. In several species of periwinkles (genus *Littorina*) and their relatives, widespread snails common to many coastal habitats, males prefer larger females under certain circumstances (Ng et al., 2013), likely for a fecundity benefit. For example, in the species *Littoraria ardouiniana* larger males prefer to follow the mucus trails of larger females (Ng and Williams, 2014). Assortative mating has been reported several times, and male preferences for larger females might be the mechanism behind this pattern (Erlandsson and Rolan-Alvarez, 1998; Pal et al., 2006; Avaca et al., 2012; Ng and Williams, 2014, Saltin et al., 2013).

Studies of mating behavior in cephalopods are not abundant. There is evidence for open and cryptic female choice (Franklin et al., 2014; Sato et al., 2014, 2017; Squires et al., 2014, 2015; Schnell et al., 2015; Morse et al., 2017), though no evidence for male mate choice. Yet, conditions like the consumption of male spermatophores in *Sepiadarium austrinum* seem to predict male mate choice. Males of the southern bottletail squid, *Sepiadarium austrinum*, were more likely to provide spermatophores for females that were larger and carried more eggs (Wegener et al., 2013).

Clitellata

Earthworms are also simultaneous hermaphrodites. In *Eisenia andrei* individuals seem to prefer larger partners and invest more sperm when a partner is more fecund (Velando et al., 2008).

Insects

Male mate choice in insects was extensively covered by Bonduriansky (2001). In his influential review he listed 58 species with some form of male mate choice. In this germane paper, he also defined key conditions influencing the evolution of male mate choice, potential mechanisms, and the relationship with sex roles (see Chapter 2). Given that there are an estimated 1 million species of insects, 58 is not even scratching the surface. The 58 examples Bonduriansky reported come from 37 families and 11 orders within the insects (Bonduriansky, 2001). Despite the haphazard collection of these examples a few patterns emerged in the reviewed taxa relative to the likely adaptive benefits of male choice. In Coleoptera (beetles), most cases of male mate choice seem to be associated with low search cost, which favors male choosiness in tight female associations. By contrast, in flies (Diptera), male choosiness seems to arise owing to large variability in female quality (the Dipteran examples are dominated by studies on *Drosophila*), whereas the driving factor in Orthoptera, Hemiptera, and Homoptera is large male investment in—among other things—spermatophores. Of course, this is painting the picture with broad strokes, and closer inspection reveals much more nuance. Perhaps more interesting are cases that are not as easily predicted by theory. In many cases, males are choosy despite high costs for mate searching and no investment in the female or the offspring. In such cases it would be good to investigate other pathways for males to develop choosiness, such as differences in female quality and sex ratio. Indeed, this pattern seems to be quite common and Bonduriansky suggests that male mate choice is widespread in butterflies (Bonduriansky, 2001).

There are some additional examples of male mate choice. Monogamous bark beetle (*Dendroctonus valens*) males show preferences for individual females based on odor. The pairs formed this way have higher reproductive fitness as compared to pairs formed with nonpreferred females (Chen et al., 2012). In another species of bark beetle, *Bolitotherus cornutus*, males preferred larger females both in a laboratory study and in the field (Formica et al., 2016). In the mealworm beetle (*Tenebrio molitor*), males show a preference for virgin and mature females by adjusting their ejaculate (Carazo et al., 2004). In this species both sexes choose (Reyes-Ramírez et al., 2020). The mechanism for this is chemosensory. A comparable finding was reported for stick insects, *Timema* (Riesch et al., 2017), and the mechanism is also chemosensory.

In a praying mantis, *Pseudomantis albofimbriata*, males did not respond to differences in female quality in an experimental approach, although they were predicted to do so (Barry et al., 2010). Another species of praying mantis, *Tenodera angustipennis*, shows male preference for well-fed females (Kadoi et al., 2017). Apparently, in this sexually cannibalistic group, this preference is not because of a fecundity benefit, but reflects a higher probability of male escape from a well-fed female after copulation. This situation seems rather similar to the one found in sexually cannibalistic spiders.

Using *Drosophila melanogaster* (Edward and Chapman, 2012) showed that male mate choice has strong fitness benefits for males. Interestingly, Long and colleagues found that male preferences and the resulting persistent mating efforts with high-quality females, can impose heavy costs for those females, and have significant evolutionary consequences via a fertility reduction for large, preferred females (Long et al., 2009).

In some species of *Drosophila*, e.g. *Drosophila bifurcata*, male sperm are very large. In these cases male investment in sperm can be almost as large as female investment in eggs (Bjork and Pitnick, 2006). This leads to intensive male mate choice with preferences for mated and larger females (Lüpold et al., 2010), also in *D. melanogaster* (Byrne and Rice, 2006). The extremely large sperm of some species are considered ornamental traits (Lüpold et al., 2016).

Another insect, a parasitoid wasp, *Spalangia endius*, shows male preferences for virgin females. Virgin and nonvirgin males did not differ in their preference (King et al., 2005). Furthermore, in a species of thrips (*Frankliniella occidentalis*), an agricultural pest, males prefer to mate with virgin females, but only when they themselves are experienced and not virgins (Akinyemi and Kirk, 2019). This points to an interesting role of learning and ontogeny in male mate preferences.

In some *Heliconius* butterflies males show mate guarding of pupae and mate with the eclosing females right after they emerge. They provide large spermatophores and also cyanide, a toxic substance that protects the eggs and the female (Cardoso et al., 2009). In pupal mating female choice is effectively undermined as a male mates with the female as soon as she emerges (Thurman et al., 2018). Male competition for pupae is intense (Deinert et al., 1994).

Another mechanism for mate choice is found in a bush cricket, *Isophya rizeensis*, where males allocate fewer sperm per spermatophore in a female-biased sex ratio. This shows again how important OSR is. And this seems to come close to a form of cryptic male mate choice (Yiğit et al., 2019). In a sex-role reversed butterfly from Africa, *Bicyclus anynana*, males choose females based on wing ornaments, and spend more time with the preferred female, likely increasing time to transfer sperm (Ng et al., 2017).

Heteropterans provide a few interesting examples of male mate choice. In giant water bugs, *Belostoma lutarium*, male mate preferences for female size have been found (Thrasher et al., 2015) (Chapter 6). In an assassin bug, *Rhinocoris tristis* (or *Rhynocoris*), male choice via differential mate guarding was reported. This is one of the few insects that provides paternal care and shows male mate choice. Finally, I want to provide another example for mutual mate choice in an insect (Chapter 9). In the stinkbug, *Acrosternum hilare*, females and males both prefer larger mates (Capone, 1995), which has been proposed as mutual mate choice. The benefit for males in this study was increased fecundity with increased female size.

Spiders

Spiders have amazingly complex mating behavior. Males are generally thought of as choosy, often preferring virgin females, and with chemical communication as a major sensory channel (Gaskett, 2007). Males use pedipalps to transfer sperm (Foelix, 2011), an interesting parallel to some cephalopods, which use specialized tentacles. In orb-weaving spiders, male mate choice might be relatively easy to evolve because males can visit females at their nest and compare them. In many species of spider males are limited to a maximum of two matings in their life. Even if most males will never mate, a limit of the maximum success seems odd to evolve. To transfer sperm, they insert their specialized copulatory organ, the pedipalp, into the female receptacle, the epigynum, where it breaks off and prevents the female from receiving additional matings. Males must insert their left pedipalp into a left receptacle and the right pedipalp into a right receptacle. Even though males invest a lot less in sperm than females in eggs, the scarcity of mating opportunities seems to predict choosiness in males. The lack of additional mating opportunities may also have been important in the evolution of widespread sexual cannibalism (Foellmer and Fairbairn, 2003, 2004, 2005). After the second copulation, a male may as well invest in his offspring by becoming food for their mother. In a series of studies on mate choice in annual orb-weaving spiders, *Zygiella x-notata*, Bel-Venner and her team found that male mate choice drives assortative mating in this species. Larger males preferred larger females, whereas smaller males prefer smaller females (Bel-Venner et al.,

2008; Venner et al., 2010). The authors betrayed their surprise at their own finding in the title of their paper, calling it "unexpected." In another species, *Micrathena gracilis*, males preferred females that were unmated (Bukowski and Christenson, 2000; Agnarsson et al., 2006). A somewhat similar pattern was found in another orb-weaving spider, *Gasteracantha cancriformis*, (Bukowski et al., 2001). In yet another orb-weaving spider, *Trichonephila clavipes*, from Brazil, males preferred larger and recently unpaired females (Pollo et al., 2019). Interestingly, sexual cannibalism has also been reported from a sex-role reversed wolf spider from Uruguay, *Allocosa brasiliensis*, and a few other sex-role reversed species (Aisenberg et al., 2011). Male mate choice for larger female body size was also documented in a jumping spider, *Phidippus clarus* (Hoefler, 2007; Hoefler et al., 2009, 2010).

In a wolf spider, *Schizocosa oreata*, males choose females based on their mating history and their foraging history, with males courting virgin females more vigorously. There is also evidence that male experience can influence male mate choice (Meyer and Uetz, 2018). In a funnel-web spider, *Agelenopsis aperta* (Riechert and Singer, 1995), a preference for older, unmated females was detected.

In a nursery web spider, *Pisaura mirabilis*, males prefer larger nuptial gifts to present to females, presumably buying longer copulations. But the preference was condition-dependent and smaller males took part of the nuptial gift for themselves (Prokop, 2019).

Less is known about mate choice in other Chelicerata. For example, one scorpion, *Centruroides vittatus*, shows female choice, with hints of male choice via differential attention to females (Nobile and Johns, 2005). Sexual cannibalism was suggested for several species, but it was argued that the observations of females eating males are due to accidents, and not akin to the sexual cannibalism found in spiders and praying mantises (Peretti et al., 1999). Horseshoe crabs, *Limulus polyphemus*, have been well studied (Brockmann, 1990), but with a mating system resembling that of explosive breeding amphibians (Brockmann, 2002) and intensive scramble competition among males, there are no signs of male mate choice.

Crustaceans

In crabs of the genus *Uca*, where males advertise to females by waving their claws, males preferred larger females, and rejected a few of the females that passed them by, all of which appeared to be unreceptive (Reading and Backwell, 2007). In this case females differed along two axes of female quality, size (reflecting fecundity), as a continuous trait, and receptivity, a categorical trait.

Unsurprisingly, mate choice is relatively well documented in Amphipods, the small crustaceans found in saltwater and in many ponds and streams. Because mate guarding is so prominent in their mating behavior, they are easily studied in this respect. The evidence points to male preferences for larger females in the European *Gammarus pulex* (Elwood and Dick, 1989; Dick and Elwood, 1990). A theoretical argument, however, predicted the opposite (Härdling et al., 2004). Males may also guard and hold on to females close to molting, potentially to minimize the time they invest in mate guarding (Ward, 1988). While male preferences seem to play a role, female resistance can also influence the patterns of mating (Bollache and Cézilly, 2004). In *Gammarus roeselii* an interesting relationship with density (and hence encounter rates between potential partners) and male choosiness exists, with the latter being reduced in low-density conditions (Lipkowski et al., 2019).

In *Hyalella azteca*, a North American species, mate guarding and pair formation were influenced by the OSR, and a female-biased OSR seems to provide males more opportunity to exercise preference. When males are more abundant, competition is more intense, and choice is diminished (Wen, 1993; Cothran, 2008).

In a crayfish, *Procambarus clarkii*, both sexes preferred larger mates, and males also preferred virgin females (Aquiloni and Gherardi, 2008).

In this chapter I reviewed many studies that report some kind of male mate choice. Yet, compared with the number of papers reporting female mate choice, this number is not very impressive. The fact that I must have missed many examples does not change this, because there are many thousands of studies focusing on female choice (Chapter 1). Clearly, this calls for more studies of male mate choice. Nonetheless, there are a few patterns that emerge from this overview. First, as predicted by theory, males seem to show choosiness when they invest in the mating interaction. This investment can range from investment in mating partners, offspring, or in the mating itself. Second, the sex ratio is important. When females are abundant and males rare, males can show choosiness. This is especially striking in systems that are dynamic and where males track the current sex ratio, which can change either locally or temporally. This is the case in gobies (Chapter 6). Furthermore, female differences in quality favor male choosiness. The most obvious trait in which females can differ is fecundity, with size as the likely correlate evaluated by males. It is highly unlikely that size is the only trait for which female variability can be detected, albeit it is clearly important. It might also be the most intuitive to look for and the easiest to study. All of these factors will be covered further in the following chapters.

3.7 Short summary

Male mate choice is actually quite common and taxonomically widespread. This is not really surprising, but what is noteworthy is that male mate choice is often discovered not because researchers were actively trying to detect it or because theory

predicted it, but as a—sometimes surprising—side effect. Nonetheless, the many examples show that male mate choice is on its way into the scientific mainstream. The majority of the studies identify female fecundity as the adaptive benefit for male mate choice and size as the trait males use to identify differences in female quality. I think, however, that we are overlooking other mechanisms. They will be discussed in the following chapters.

3.8 Additional reading

There are not many systematic reviews of male mate choice, but they are worth reading. This is in part to see how the treatment of underlying theory has changed (Sargent et al., 1986; Krupa, 1995; Bondurianksy, 2001; Schlupp, 2018).

3.9 References

AGNARSSON, I., AVILES, L., CODDINGTON, J. A., & MADDISON, W. P. 2006. Sociality in Theridiid spiders: repeated origins of an evolutionary dead end. *Evolution*, 60, 2342–51.

AISENBERG, A., COSTA, F. G., & GONZÁLEZ, M. 2011. Male sexual cannibalism in a sand-dwelling wolf spider with sex role reversal. *Biological Journal of the Linnean Society*, 103, 68–75.

Table 3.1

Species	Taxon	Male preference for	Putative benefit
Isophya rizeensis	*Insect*	*Sex-ratio dependent*	*Avoid cost for sperm production*
Campylomormyrus compressirostris	*Fish*	*Conspecific*	*Avoiding heterospecific matings*
Cyprinodon sp.	*Fish*	*Conspecific*	*Avoiding heterospecifc matings*
Poecilia latipinna	*Fish*	*Conspecific*	*Avoiding heterospecific matings*
Poecilia mexicana	*Fish*	*Conspecific*	*Avoiding heterospecific matings*
Poeciliopsis sp.	*Fish*	*Conspecific*	*Avoiding heterospecific matings*
Spermophilus tridecemlineatus	*Mammal*	*Size*	*Avoiding sperm competition*
Lonchura striata	*Bird*	*Individual females*	*Best fit?*
Pisaura mirabilis	*Spider*	*Larger nuptial gifts*	*Copulation time*
Tenodera angustipennis	*Insect*	*Well-fed partner*	*Direct benefit*
Mus musculus	*Mammal*	*Freeness of parasites*	*Direct benefit*
Sceloporus undulatus	*Reptile*	*Body condition*	*Direct benefit*

Species	Taxon	Male preference for	Putative benefit
Oncorhynchus kisutch	Fish	Size, color	Fecundity
Moniliformis moniliformis	Acathocephala	Size	Fecundity
Bufo andrewsi	Amphibian	Size	Fecundity
Desmognathus fuscus	Amphibian	Size	Fecundity
Desmognathus ochrophaeus	Amphibian	Size	Fecundity
Desmognathus santeelah	Amphibian	Size	Fecundity
Notophthalmus viridescens	Amphibian	Size	Fecundity
Rana sylvatica (Lithobates sylvaticus)	Amphibian	Size	Fecundity
Triturus vulgaris	Amphibian	Size	Fecundity
Gymnorhinus cyanocephalus	Bird	Weight	Fecundity
Taniopygia guttata	Bird	Size	Fecundity
Eisenia andrei	Clitellata	Size	Fecundity
Gammarus pulex	Crustacean	Size	Fecundity
Gammarus roeselii	Crustacean	Size	Fecundity
Uca	Crustacean	Size	Fecundity
Astatotilapia flaviijosephi	Fish	Size	Fecundity
Austrolebias reicherti	Fish	Size	Fecundity
Cichlasoma nigrofasciatum	Fish	Size	Fecundity
Etheostoma caeruleum	Fish	Size	Fecundity
Gambusia holbrooki	Fish	Size	Fecundity
Gobiusculus flavescens	Fish	Size	Fecundity
Ophioblennius atlanticus	Fish	Size, age	Fecundity
Oryzias latipes	Fish	More eggs	Fecundity
Pelvicachromis taeniatus	Fish	Size, relatedness	Fecundity
Poecilia reticulata	Fish	Size	Fecundity
Pseudomugil signifer	Fish	Size	Fecundity
Sarotherodon galilaeus	Fish	Size	Fecundity
Stegastes leucosticus	Fish	Size	Fecundity
Thalassoma bifasciatum	Fish	Size	Fecundity
Acrosternum hilare	Insect	Size	Fecundity
Belostoma lutarium	Insect	Size	Fecundity
Bolitotherus cornutus	Insect	Size	Fecundity
Drosophila bifurcata	Insect	Size	Fecundity
Drosophila melanogaster	Insect	Size	Fecundity
Rhinocoris tristis	Insect	Size	Fecundity
Bison bison	Mammal	Reproductive potential	Fecundity
Homo sapiens	Mammal	Size and other	Fecundity
Lemur catta	Mammal	Age	Fecundity
Megaptera novaeangliae	Mammal	Size	Fecundity
Ovis aries	Mammal	Size	Fecundity

(continued)

Table 3.1 Continued

Species	Taxon	Male preference for	Putative benefit
Pan troglodytes	Mammal	Older age	Fecundity
Littoraria ardouiniana	Mollusc	Size	Fecundity
Sepiadarium austrinum	Mollusc	Size	Fecundity
Lacerta agilis	Reptile	Size	Fecundity
Platysaurus broadleyi	Reptile	Size	Fecundity
Thamnophis sirtalis	Reptile	Size	Fecundity
Phidippus clarus	Spider	Size	Fecundity
Xiphophorus malinche	Fish	Size	Fecundity
Gasterosteus aculeatus	Fish	Size and ornaments	Fecundity and readiness
Ichthyosaura alpestris	Amphibian	Color	Fecundity?
Passer domesticus	Bird	Older females	Fecundity?
Hyalella azteca	Crustacean	Density dependent	Fecundity?
Canis lupus	Mammal	Social rank	Fecundity?
Crocuta crocuta	Mammal	Rank	Fecundity?
Zygiella x-notata	Spider	Assortative	Fecundity?
Agelenopsis aperta	Spider	Older or unmated	Fecundity/Paternity assurance
Trichonephila clavipes	Spider	Size/unmated	Fecundity/paternity assurance
Junco hyemalis	Bird	Population	Hybrid avoidance
Betta splendens	Fish	Conspecific	Hybridization avoidance?
Luscinia svecica	Bird	Bigger ornament	Indicator for fecundity?
Crotophytus collaris	Reptile	Ornament	Indicator for fecundity?
Ctenophorus maculosus	Reptile	Ornament	Indicator for fecundity?
Parus major	Bird	Color ornament	Juvenile survival
Tenodera angustipennis	Insect	Well fed females	Male escape
Kryptolebias marmoratus	Fish	MHC	Matching partner
Equus ferus caballus	Mammal	MHC	Matching partner
Mandrillus sphinx	Mammal	MHC	Matching partner
Procyon lotor	Mammal	MHC	Matching partner
Dendroctonus valens	Insect	Odor	Matching partner?
Limulus polyphenus	Chelicerate	No choice	No benefit
Brachyraphis rhabdophora	Fish	No choice	No benefit
Carcharhinus sorrah	Fish	None found	No benefit
Percina roanoka	Fish	No preference	No benefit
Pseudomantis albofimbriata	Insect	No choice	No benefit
Etheostoma spectabile	Fish	None	None
Rana esculenta	Amphibian	No choice	No discrimination
Rana lessonae	Amphibian	No choice	No discrimination
Carpodacus mexicanus	Bird	Older/color	Not clear
Culaea inconstans	Fish	Color	Not clear
Dermogynis colletti	Fish	Ornament	Not clear

Species	Taxon	Male preference for	Putative benefit
Gazella marica	*Mammal*	*Not clear*	*Not clear*
Oreamnos americanus	*Mammal*	*Age*	*Not clear*
Mercurialis annua	*Plant*	*Not clear*	*Not clear*
Sceloperus virgatus	*Reptile*	*Color ornament*	*Not clear*
Aethia cristatella	*Bird*	*Enhanced ornament*	*Novelty?*
Lonchura leucogastroides	*Bird*	*New ornament*	*Novelty*
Mungus mungo	*Mammal*	*Relatedness*	*Outbreeding*
Anolis carolinensis	*Reptile*	*Familiarity*	*Outbreeding*
Anolis sagrei	*Reptile*	*Familiarity*	*Outbreeding?*
Erythrura gouldiae	*Bird*	*Morph*	*Own morph*
Symphodus ocellatus	*Fish*	*Social context*	*Paternity assurance*
Frankliniella occidentalis	*Insect*	*Virgin females*	*Paternity assurance*
Heliconius spec.	*Insect*	*Virgin females*	*Paternity assurance*
Spalangia endius	*Insect*	*Virgin females*	*Paternity assurance*
Tenebrio molitor	*Insect*	*Virgin females*	*Paternity assurance*
Gasteracantha cancriformis	*Spider*	*Virgin females*	*Paternity assurance*
Micrathena gracilis	*Spider*	*Virgin females*	*Paternity assurance*
Schizocosa oreata	*Spider*	*Virgin female*	*Paternity assurance*
Puntius titteya	*Fish*	*Coloration*	*Quality indicator*
Neochromis omnicaeruleus	*Fish*	*Color morph*	*Speciation?*
Alytes cisternasii	*Amphibian*	*Calls*	*Unclear*
Alytes muletensis	*Amphibian*	*Calls*	*Unclear*
Brachionus plicatilis	*Rotatorian*	*Age*	*Young females*

AKINYEMI, A. O. & KIRK, W. D. J. 2019. Experienced males recognise and avoid mating with non-virgin females in the western flower thrips. *PLoS ONE*, 14, e0224115.

ALMADA, V. C., GONÇALVES, E. J., OLIVEIRA, R. F., & SANTOS, A. J. 1995. Courting females: ecological constraints affect sex roles in a natural population of the blenniid fish *Salaria pavo*. *Animal Behaviour*, 49, 1125–7.

ALMADA, V. C., GONÇALVES, E. J., SANTOS, A. J., & BAPTISTA, C. 1994. Breeding ecology and nest aggregations in a population of *Salaria pavo* (Pisces: Blenniidae) in an area where nest sites are very scarce. *Journal of Fish Biology*, 45, 819–30.

ALMOJIL, D. 2020. Male mate choice in the spot-tail shark *Carcharhinus sorrah*: are males choosy or opportunistic? *Journal of Negative Results*, 13, 1–11.

ALONZO, S. H. & WARNER, R. R. 1999. A trade-off generated by sexual conflict: Mediterranean wrasse males refuse present mates to increase future success. *Behavioral Ecology*, 10, 105–11.

AMUNDSEN, T. 2018. Sex roles and sexual selection: lessons from a dynamic model system. *Current Zoology*, 64, 363–92.

AMUNDSEN, T., FORSGREN, E., & HANSEN, L. T. 1997. On the function of female ornaments: male bluethroats prefer colourful females. *Proceedings of the Royal Society of London—Series B: Biological Sciences*, 264, 1579–86.

ANDERSSON, M. 1994. *Sexual Selection*. Princeton, NJ: Princeton University Press.

ANTHES, N., WERMINGHAUSEN, J., & LANGE, R. 2014. Large donors transfer more sperm, but depletion is faster in a promiscuous hermaphrodite. *Behavioral Ecology and Sociobiology*, 68, 477–83.

AQUILONI, L. & GHERARDI, F. 2008. Mutual mate choice in crayfish: large body size is selected by both sexes, virginity by males only. *Journal of Zoology*, 274, 171–9.

ARNOLD, S. J. 1994a. Bateman's principles and the measurement of sexual selection in plants and animals. *The American Naturalist*, 144, S126–49.

ARNOLD, S. J. 1994b. Is there a unifying concept of sexual selection that applies to both plants and animals? *The American Naturalist,* 144, S1–12.

AULD, H. L., JESWIET, S. B., & GODIN, J. G. J. 2015. Do male Trinidadian guppies adjust their alternative mating tactics in the presence of a rival male audience? *Behavioral Ecology and Sociobiology,* 69, 1191–9.

AULD, H. L., RAMNARINE, I. W., & GODIN, J. G. J. 2017. Male mate choice in the Trinidadian guppy is influenced by the phenotype of audience sexual rivals. *Behavioral Ecology,* 28, 362–72.

AVACA, M. S., NARVARTE, M., & MARTIN, P. 2012. Size-assortative mating and effect of maternal body size on the reproductive output of the nassariid *Buccinanops globulosus. Journal of Sea Research,* 69, 16–22.

BAIRD, T. A. 2004. Reproductive coloration in female collared lizards, *Crotophytus collaris,* stimulates courtship by males. *Herpetologica,* 60, 337–48.

BAKKER, T. C. M. & ROWLAND, W. J. 1995. Male mating preference in sticklebacks: effects of repeated testing and own attractiveness. *Behaviour,* 132, 935–49.

BALDAUF, S. A., BAKKER, T. C. M., HERDER, F., KULLMANN, H., & THÜNKEN, T. 2010. Male mate choice scales female ornament allometry in a cichlid fish. *BMC Evolutionary Biology,* 10, 310. https://doi.org/10.1186/1471-2148-10-301.

BALDAUF, S. A., BAKKER, T. C. M., KULLMANN, H., & THÜNKEN, T. 2011. Female nuptial coloration and its adaptive significance in a mutual mate choice system. *Behavioral Ecology,* 22, 478–85.

BALSHINE-EARN, S. 1996. Reproductive rates, operational sex ratios and mate choice in St. Peter's fish. *Behavioral Ecology and Sociobiology,* 39, 107–16.

BARRY, K. L., HOLWELL, G. I., & HERBERSTEIN, M. E. 2010. Multimodal mate assessment by male praying mantids in a sexually cannibalistic mating system. *Animal Behaviour,* 79, 1165–72.

BARRY, K. L. & KOKKO, H. 2010. Male mate choice: why sequential choice can make its evolution difficult. *Animal Behaviour,* 80, 163–9.

BASOLO, A. L. 2004. Variation between and within the sexes in body size preferences. *Animal Behaviour,* 68, 75–82.

BATEMAN, A. J. 1948. Intra-sexual selection in *Drosophila. Heredity,* 2, 349–68.

BEECHING, S. C. & HOPP, A. B. 1999. Male mate preference and size-assortative pairing in the convict cichlid. *Journal of Fish Biology,* 55, 1001–8.

BEL-VENNER, M. C., DRAY, S., ALLAINE, D., MENU, F., & VENNER, S. 2008. Unexpected male choosiness for mates in a spider. *Proceedings of the Royal Society of London—Series B: Biological Sciences,* 275, 77–82.

BERGER, J. 1989. Female reproductive potential and its apparent evaluation by male mammals. *Journal of Mammalogy,* 70, 347–58.

BERVEN, K. A. 1981. Mate choice in the wood frog, *Rana sylvatica. Evolution,* 35, 707–22.

BETANCUR-R, R., BROUGHTON, R. E., WILEY, E. O., CARPENTER, K., LÓPEZ, J. A., LI, C., HOLCROFT, N. I., ARCILA, D., SANCIANGCO, M., & CURETON II, J. C. 2013. The tree of life and a new classification of bony fishes. *PLoS Currents,* 5.

BETTS, M. G., HADLEY, A. S., & KRESS, W. J. 2015. Pollinator recognition by a keystone tropical plant. *Proceedings of the National Academy of Sciences,* 112, 3433–8.

BJORK, A. & PITNICK, S. 2006. Intensity of sexual selection along the anisogamy–isogamy continuum. *Nature,* 441, 742–5.

BLOCH, A., ESTELA, V., LEESE, J., & ITZKOWITZ, M. 2016. Male mate preference and size-assortative mating in convict cichlids: a role for female aggression? *Behavioural Processes,* 130, 81–5.

BOLLACHE, L. & CÉZILLY, F. 2004. Sexual selection on male body size and assortative pairing in *Gammarus pulex* (Crustacea: Amphipoda): field surveys and laboratory experiments. *Journal of Zoology,* 264, 135–41.

BONDURIANSKY, R. 2001. The evolution of male mate choice in insects: a synthesis of ideas and evidence. *Biological Reviews,* 76, 305–39.

BROCKMANN, H. J. 1990. Mating behavior of horseshoe crabs, *Limulus polyphemus. Behaviour,* 114, 206–20.

BROCKMANN, H. J. 2002. An experimental approach to altering mating tactics in male horseshoe crabs (*Limulus polyphemus*). *Behavioral Ecology,* 13, 232–8.

BUKOWSKI, T. C. & CHRISTENSON, T. E. 2000. Determinants of mating frequency in the spiny orbweaving spider, *Micrathena gracilis* (Araneae: Araneidae). *Journal of Insect Behavior,* 13, 331–52.

BUKOWSKI, T. C., LINN, C. D., & CHRISTENSON, T. E. 2001. Copulation and sperm release in *Gasteracantha cancriformis* (Araneae: Araneidae): differential male behaviour based on female mating history. *Animal Behaviour,* 62, 887–95.

BURD, M. 1994. Bateman's principle and plant reproduction: the role of pollen limitation in fruit and seed set. *The Botanical Review,* 60, 83–139.

BUSH, S. L. & BELL, D. J. 1997. Courtship and female competition in the Majorcan midwife toad, *Alytes muletensis. Ethology,* 103, 292–303.

BUSH, S. L., DYSON, M. L., & HALLIDAY, T. R. 1996. Selective phonotaxis by males in the Majorcan midwife toad. *Proceedings of the Royal Society of London—Series B: Biological Sciences,* 263, 913–17.

BUSS, D. 2015. *Evolutionary Psychology.* Abingdon: Routledge.

BUSS, D. M. & SCHMITT, D. P. 2019. Mate preferences and their behavioral manifestations. *Annual Review of Psychology,* 70, 77–110.

BYRNE, P. G. & RICE, W. R. 2006. Evidence for adaptive male mate choice in the fruit fly *Drosophila melanogaster*. *Proceedings of the Royal Society of London—Series B: Biological Sciences*, 273, 917–22.

CANDOLIN, U. & SALESTO, T. 2009. Does competition allow male mate choosiness in threespine sticklebacks? *American Naturalist*, 173, 273–7.

CAPONE, T. A. 1995. Mutual preference for large mates in green stink bugs, *Acrosternum hilare* (Hemiptera, Pentatomidae). *Animal Behaviour*, 49, 1335–44.

CARAZO, P., SANCHEZ, E., FONT, E., & DESFILIS, E. 2004. Chemosensory cues allow male *Tenebrio molitor* beetles to assess the reproductive status of potential mates. *Animal Behaviour*, 68, 123–9.

CARDOSO, M. Z., ROPER, J. J., & GILBERT, L. E. 2009. Prenuptial agreements: mating frequency predicts gift-giving in *Heliconius* species. *Entomologia Experimentalis et Applicata*, 131, 109–14.

CARUSO, C. M., EISEN, K. E., MARTIN, R. A., & SLETVOLD, N. 2019. A meta-analysis of the agents of selection on floral traits. *Evolution*, 73, 4–14.

CASPERS, B. & WITTE, K. 2006. Sexual imprinting on a novel blue ornament in zebra finches. *Behaviour*, 143, 969–91.

CHAN, R., STUART-FOX, D., & JESSOP, T. S. 2009. Why are females ornamented? A test of the courtship stimulation and courtship rejection hypotheses. *Behavioral Ecology*, 20, 1334–42.

CHEN, B.-J., LIU, K., ZHOU, L.-J., GOMES-SILVA, G., SOMMER-TREMBO, C., & PLATH, M. 2018. Personality differentially affects individual mate choice decisions in female and male western mosquitofish (*Gambusia affinis*). *PLoS ONE*, 13, e0197197.

CHEN, H.-F., SALCEDO, C., & SUN, J.-H. 2012. Male mate choice by chemical cues leads to higher reproductive success in a bark beetle. *Animal Behaviour*, 83, 421–7.

CICCOTTO, P. J., GUMM, J. M., & MENDELSON, T. C. 2014. No evidence for color or size preference in either sex of a dichromatic stream fish, *Percina roanoka*. *Environmental Biology of Fishes*, 97, 187–95.

CLUTTON-BROCK, T. 2007. Sexual selection in males and females. *Science*, 318, 1882–5.

CLUTTON-BROCK, T. 2016. *Mammal Societies*. Hoboken, NJ: John Wiley & Sons.

COCUCCI, A. A., MARINO, S., BARANZELLI, M., WIEMER, A. P., & SÉRSIC, A. 2014. The buck in the milkweed: evidence of male–male interference among pollinaria on pollinators. *New Phytologist*, 203, 280–6.

CONFER, J. C., EASTON, J. A., FLEISCHMAN, D. S., GOETZ, C. D., LEWIS, D. M., PERILLOUX, C., & BUSS, D. M. 2010. Evolutionary psychology: controversies, questions, prospects, and limitations. *American Psychologist*, 65, 110.

CONROY-BEAM, D. & BUSS, D. M. 2016. Do mate preferences influence actual mating decisions? Evidence from computer simulations and three studies of mated couples. *Journal of Personality and Social Psychology*, 111, 53–66.

CONROY-BEAM, D., BUSS, D. M., PHAM, M. N., & SHACKELFORD, T. K. 2015. How sexually dimorphic are human mate preferences? *Personality and Social Psychology Bulletin*, 41, 1082–93.

CÔTÉ, I. & HUNTE, W. 1989. Male and female mate choice in the redlip blenny: why bigger is better. *Animal Behaviour*, 38, 78–88.

CÔTÉ, I. & HUNTE, W. 1993. Female redlip blennies prefer older males. *Animal Behaviour*, 46, 203–5.

COTHRAN, R. D. 2008. The mechanistic basis of a large male mating advantage in two freshwater amphipod species. *Ethology*, 114, 1145–53.

CRAIG, A. S., HERMAN, L. M., & PACK, A. A. 2002. Male mate choice and male-male competition coexist in the humpback whale (*Megaptera novaeangliae*). *Canadian Journal of Zoology—Revue Canadienne De Zoologie*, 80, 745–55.

CULUMBER, Z. W., ANAYA-ROJAS, J. M., BOOKER, W. W., HOOKS, A. P., LANGE, E. C., PLUER, B., RAMÍREZ-BULLÓN, N., & TRAVIS, J. 2019. Widespread biases in ecological and evolutionary studies. *BioScience*, 69, 631–40.

DAWKINS, R. 1976. *The Selfish Gene*. Oxford: Oxford University Press.

DEINERT, E., LONGINO, J., & GILBERT, L. E. 1994. Mate competition in butterflies. *Nature*, 370, 23.

DELLINGER, M., ZHANG, W., BELL, A. M., & HELLMANN, J. K. 2018. Do male sticklebacks use visual and/or olfactory cues to assess a potential mate's history with predation risk? *Animal Behaviour*, 145, 151–9.

DICK, J. T. A. & ELWOOD, R. W. 1990. Symmetrical assessment of female quality by male *Gammarus pulex* (Amphipoda) during struggles over precopula females. *Animal Behaviour*, 40, 877–83.

DORKEN, M. & PERRY, L. 2017. Correlated paternity measures mate monopolization and scales with the magnitude of sexual selection. *Journal of Evolutionary Biology*, 30, 377–87.

DOSEN, L. D. & MONTGOMERIE, R. 2004a. Female size influences mate preferences of male guppies. *Ethology*, 110, 245–55.

DOSEN, L. D. & MONTGOMERIE, R. 2004b. Mate preferences by male guppies (*Poecilia reticulata*) in relation to the risk of sperm competition. *Behavioral Ecology and Sociobiology*, 55, 266–71.

DOUGHERTY, L. R. & SHUKER, D. M. 2014. The effect of experimental design on the measurement of mate choice: a meta-analysis. *Behavioral Ecology*, 26, 311–19.

DUGAS, M. B., WAMELINK, C. N., KILLIUS, A. M., & RICHARDS-ZAWACKI, C. L. 2016. Parental care is beneficial for offspring, costly for mothers, and limited by family size in an egg-feeding frog. *Behavioral Ecology*, 27, 476–83.

DYSON, M. L., BUSH, S. L., & HALLIDAY, T. R. 1998. Phonotaxis by female Majorcan midwife toads, *Alytes muletensis*. *Behaviour*, 135, 213–30.

EAST, M. L. & HOFER, H. 2001. Male spotted hyenas (*Crocuta crocuta*) queue for status in social groups dominated by females. *Behavioral Ecology*, 12, 558–68.

EDWARD, D. A. & CHAPMAN, T. 2012. Measuring the fitness benefits of male mate choice in *Drosophila melanogaster*. *Evolution: International Journal of Organic Evolution*, 66, 2646–53.

ELLISON, A., JONES, J., INCHLEY, C., & CONSUEGRA, S. 2013. Choosy males could help explain androdioecy in a selfing fish. *American Naturalist*, 181, 855–62.

ELWOOD, R. W. & DICK, J. T. A. 1989. Assessments and decisions during mate choice in *Gammarus pulex* (Amphipoda). *Behaviour*, 109, 235.

ENGELER, B. & REYER, H.-U. 2001. Choosy females and indiscriminate males: mate choice in mixed populations of sexual and hybridogenetic water frogs (*Rana lessonae*, *Rana esculenta*). *Behavioral Ecology*, 12, 600–6.

ERLANDSSON, J. & ROLAN-ALVAREZ, E. 1998. Sexual selection and assortative mating by size and their roles in the maintenance of a polymorphism in Swedish *Littorina saxatilis* populations. *Hydrobiologia*, 378, 59–69.

FISCHER, E. K., ROLAND, A. B., MOSKOWITZ, N. A., VIDOUDEZ, C., RANAIVORAZO, N., TAPIA, E. E., TRAUGER, S. A., VENCES, M., COLOMA, L. A., & O'CONNELL, L. A. 2019. Mechanisms of convergent egg provisioning in poison frogs. *Current Biology*, 29, 4145–51.

FISKE, P. & AMUNDSEN, T. 1997. Female bluethroats prefer males with symmetric colour bands. *Animal Behaviour*, 54, 81–7.

FOELIX, R. 2011. *Biology of Spiders*. Cary, NC: Oxford University Press.

FOELLMER, M. W. & FAIRBAIRN, D. J. 2003. Spontaneous male death during copulation in an orb-weaving spider. *Proceedings of the Royal Society of London—Series B: Biological Sciences (Supplement)*, 270, S183–5.

FOELLMER, M. W. & FAIRBAIRN, D. J. 2004. Males under attack: sexual cannibalism and its consequences for male morphology and behaviour in an orb-weaving spider. *Evolutionary Ecology Research*, 6, 163–81.

FOELLMER, M. W. & FAIRBAIRN, D. J. 2005. Competing dwarf males: sexual selection in an orb-weaving spider. *Journal of Evolutionary Biology*, 18, 629–41.

FOOTE, C. J. 1988. Male mate choice dependent on male size in salmon. *Behaviour*, 106, 63–80.

FOOTE, C. J., BROWN, G. S., & HAWRYSHYN, C. W. 2004. Female colour and male choice in sockeye salmon: implications for the phenotypic convergence of anadromous and nonanadromous morphs. *Animal Behaviour*, 67, 69–83.

FOOTE, C. J. & LARKIN, P. A. 1988. The role of male choice in the assortative mating of anadromous and non-anadromous sockeye salmon (*Oncorhynchus nerka*). *Behaviour*, 106, 43–62.

FORMICA, V. A., DONALD-CANNON, H., & PERKINS-TAYLOR, I. E. 2016. Consistent patterns of male mate preference in the laboratory and field. *Behavioral Ecology and Sociobiology*, 70, 1805–12.

FRANKLIN, A. M., SQUIRES, Z. E., & STUART-FOX, D. 2014. Does predation risk affect mating behavior? An experimental test in dumpling squid (*Euprymna tasmanica*). *PLoS ONE*, 9.

FUKUDA, S. & KARINO, K. 2014. Male red coloration, female mate preference, and sperm longevity in the cyprinid fish *Puntius titteya*. *Environmental Biology of Fishes*, 97, 1197–205.

GANEM, G., LITEL, C., & LENORMAND, T. 2008. Variation in mate preference across a house mouse hybrid zone. *Heredity*, 100, 594.

GASKETT, A. 2007. Spider sex pheromones: emission, reception, structures, and functions. *Biological Reviews*, 82, 27–48.

GASKETT, A. C. 2011. Orchid pollination by sexual deception: pollinator perspectives. *Biological Reviews*, 86, 33–75.

GIRNDT, A., COCKBURN, G., SÁNCHEZ-TÓJAR, A., HERTEL, M., BURKE, T., & SCHROEDER, J. 2019. Male age and its association with reproductive traits in captive and wild house sparrows. *Journal of Evolutionary Biology*, 32, 1432–43.

GÓMEZ, A. & SERRA, M. 1996. Mate choice in male *Brachionus plicatilis* rotifers. *Functional Ecology*, 10, 681–7.

GOTELLI, N. J. & MOORE, J. 1992. Altered host behavior in a cockroach acanthocephalan association. *Animal Behaviour*, 43, 949–59.

GOURBAL, B. E. F. & GABRION, C. 2004. A study of mate choice in mice with experimental *Taenia crassiceps* cysticercosis: can males choose? *Canadian Journal of Zoology*, 82, 635–43.

GOYMANN, W., MOORE, I. T., & OLIVEIRA, R. F. 2019. Challenge hypothesis 2.0: a fresh look at an established idea. *BioScience*, 69, 432–42.

GRANT, J., CASEY, P., BRYANT, M., & SHAHSAVARANI, A. 1995. Mate choice by male Japanese medaka (Pisces, Oryziidae). *Animal Behaviour*, 50, 1425–8.

HANSEN, L. T., AMUNDSEN, T., & FORSGREN, E. 1999. Symmetry: attractive not only to females. *Proceedings of the Royal Society of London—Series B: Biological Sciences*, 266, 1235–40.

HÄRDLING, R., KOKKO, H., & ELWOOD, R. W. 2004. Priority versus brute force: when should males begin guarding resources? *American Naturalist,* 163, 240–52.

HEAD, M. L., JACOMB, F., VEGA-TREJO, R., & JENNIONS, M. D. 2015. Male mate choice and insemination success under simultaneous versus sequential choice conditions. *Animal Behaviour,* 103, 99–105.

HEINIG, A., PANT, S., DUNNING, J. L., BASS, A., COBURN, Z., & PRATHER, J. F. 2014. Male mate preferences in mutual mate choice: finches modulate their songs across and within male-female interactions. *Animal Behaviour,* 97, 1–12.

HERDMAN, E. J. E., KELLY, C. D., & GODIN, J.-G. J. 2004. Male mate choice in the guppy (*Poecilia reticulata*): do males prefer larger females as mates? *Ethology,* 110, 97–111.

HETTYEY, A., TÖRÖK, J., & HÉVIZI, G. 2005. Male mate choice lacking in the agile frog, *Rana dalmatina. Copeia,* 2005, 403–8.

HEUBEL, K. 2018. Female mating competition alters female mating preferences in common gobies. *Current Zoology,* 64, 351–61.

HILL, G. E. 1993. Male mate choice and the evolution of female plumage coloration in the house finch. *Evolution,* 47, 1515–25.

HOEFLER, C. D. 2007. Male mate choice and size-assortative pairing in a jumping spider, *Phidippus clarus. Animal Behaviour,* 73, 943–54.

HOEFLER, C. D., CARLASCIO, A. L., PERSONS, M. H., & RYPSTRA, A. L. 2009. Male courtship repeatability and potential indirect genetic benefits in a wolf spider. *Animal Behaviour,* 78, 183–8.

HOEFLER, C. D., MOORE, J. A., REYNOLDS, K. T., & RYPSTRA, A. L. 2010. The effect of experience on male courtship and mating behaviors in a cellar spider. *American Midland Naturalist,* 163, 255–68.

HOLEKAMP, K. E., SMITH, J. E., STRELIOFF, C. C., VAN HORN, R. C., & WATTS, H. E. 2012. Society, demography and genetic structure in the spotted hyena. *Molecular Ecology,* 21, 613–32.

HONE, D. W., NAISH, D., & CUTHILL, I. C. 2012. Does mutual sexual selection explain the evolution of head crests in pterosaurs and dinosaurs? *Lethaia,* 45, 139–56.

HOYSAK, D. J. & GODIN, J.-G. J. 2007. Repeatability of male mate choice in the mosquitofish, *Gambusia holbrooki. Ethology,* 113, 1007–18.

ITZKOWITZ, M., DRAUD, M., BARNES, J., & HALEY, M. 1998. Does it matter that male beaugregory damselfish have a mate preference? *Behavioral Ecology and Sociobiology,* 42, 149–55.

JEANNERAT, E., JANETT, F., SIEME, H., WEDEKIND, C., & BURGER, D. 2017. Quality of seminal fluids varies with type of stimulus at ejaculation. *Scientific Reports,* 7, 44339.

JEANNERAT, E., MARTI, E., BERNEY, C., JANETT, F., BOLLWEIN, H., SIEME, H., BURGER, D., & WEDEKIND, C. 2018. Stallion semen quality depends on major histocompatibility complex matching to teaser mare. *Molecular Ecology,* 27, 1025–35.

JOHNSON, K. 1988a. Sexual selection in pinyon jays I: female choice and male-male competition. *Animal Behaviour,* 36, 1038–47.

JOHNSON, K. 1988b. Sexual selection in pinyon jays II: male choice and female-female competition. *Animal Behaviour,* 36, 1048–53.

JONES, A. G. & AVISE, J. C. 2001. Mating systems and sexual selection in male-pregnant pipefishes and seahorses: insights from microsatellite-based studies of maternity. *Journal of Heredity,* 92, 150–8.

JONES, I. L. & HUNTER, F. M. 1993. Mutual sexual selection in a monogamous seabird. *Nature,* 362, 238–9.

JUSTUS, K. T. & MENDELSON, T. C. 2018. Male preference for conspecific mates is stronger than females' in *Betta splendens. Behavioural Processes,* 151, 6–10.

KADOI, M., MORIMOTO, K., & TAKAMI, Y. 2017. Male mate choice in a sexually cannibalistic species: male escapes from hungry females in the praying mantid *Tenodera angustipennis. Journal of Ethology,* 35, 177–85.

KAPPELER, P. M. 2012. Mate choice. *The Evolution of Primate Societies,* 343, 366.

KEAGY, J., MINTER, R., & TINGHITELLA, R. M. 2019. Sex differences in cognition and their relationship to male mate choice. *Current Zoology,* 65, 285–93.

KEDDY-HECTOR, A. C. 1992. Mate choice in non-human primates. *American Zoologist,* 32, 62–70.

KEEGAN-ROGERS, V. 1984. Unfamiliar female mating advantage among clones of unisexual fish (Poeciliopsis: Poeciliidae). *Copeia,* 1984, 169–74.

KIMMITT, A. A., DIETZ, S. L., REICHARD, D. G., & KETTERSON, E. D. 2018. Male courtship preference during seasonal sympatry may maintain population divergence. *Ecology and Evolution,* 8, 11833–41.

KING, B., SAPORITO, K., ELLISON, J., & BRATZKE, R. 2005. Unattractiveness of mated females to males in the parasitoid wasp *Spalangia endius. Behavioral Ecology and Sociobiology,* 57, 350–6.

KNELL, R. J., NAISH, D., TOMKINS, J. L., & HONE, D. W. 2013. Sexual selection in prehistoric animals: detection and implications. *Trends in Ecology & Evolution,* 28, 38–47.

KNIEL, N., BENDER, S., & WITTE, K. 2016. Sex-specific audience effect in the context of mate choice in zebra finches. *PLoS ONE,* 11.

KOMDEUR, J., OOREBEEK, M., VAN OVERVELD, T., & CUTHILL, I. C. 2005. Mutual ornamentation, age, and reproductive performance in the European starling. *Behavioral Ecology,* 16, 805–17.

KRAAK, S. B. M. & BAKKER, T. C. M. 1998. Mutual mate choice in sticklebacks: attractive males choose big females, which lay big eggs. *Animal Behaviour,* 56, 859–66.

KRUPA, J. J. 1995. How likely is male mate choice among anurans. *Behaviour,* 132, 643–64.

LAWLOR, B. J., READ, A. F., KEYMER, A. E., PARVEEN, G., & CROMPTON, D. 1990. Non-random mating in a parasitic worm: mate choice by males? *Animal Behaviour,* 40, 870–6.

LEA, J., HALLIDAY, T. R., & DYSON, M. 2003. The mating strategy of *Alytes muletensis*: some males are less ready to mate than females. *Amphibia-Reptilia,* 24, 169–80.

LEMASTER, M. P. & MASON, R. T. 2002. Variation in a female sexual attractiveness pheromone controls male mate choice in garter snakes. *Journal of Chemical Ecology,* 28, 1269–85.

LEONARD, J. L. 2006. Sexual selection: lessons from hermaphrodite mating systems. *Integrative and Comparative Biology,* 46, 349–67.

LIAO, W. B. & LU, X. 2009. Male mate choice in the Andrew's toad *Bufo andrewsi*: a preference for larger females. *Journal of Ethology,* 27, 413–17.

LIPKOWSKI, K., PLATH, M., KLAUS, S., & SOMMER-TREMBO, C. 2019. Population density affects male mate choosiness and morphology in the mate-guarding amphipod *Gammarus roeselii* (Crustacea: Amphipoda). *Biological Journal of the Linnean Society,* 126, 899–911.

LONG, T. A., PISCHEDDA, A., STEWART, A. D., & RICE, W. R. 2009. A cost of sexual attractiveness to high-fitness females. *PLoS Biology,* 7, e1000254.

LÜDTKE, D. U. & FOERSTER, K. 2018. Choosy males court both large, colourful females and less colourful but responsive females for longer. *Animal Behaviour,* 146, 1–11.

LÜPOLD, S., MANIER, M. K., ALA-HONKOLA, O., BELOTE, J. M., & PITNICK, S. 2010. Male *Drosophila melanogaster* adjust ejaculate size based on female mating status, fecundity, and age. *Behavioral Ecology,* 22, 184–91.

LÜPOLD, S., MANIER, M. K., PUNIAMOORTHY, N., SCHOFF, C., STARMER, W. T., LUEPOLD, S. H. B., BELOTE, J. M., & PITNICK, S. 2016. How sexual selection can drive the evolution of costly sperm ornamentation. *Nature,* 533, 535–8.

MADDISON, D. R. & SCHULZ, K.-S. 2007. The Tree of Life Web Project. http://tolweb.org/tree/ (accessed November 28, 2020).

MAINGUY, J. & COTE, S. D. 2008. Age- and state-dependent reproductive effort in male mountain goats, *Oreamnos americanus. Behavioral Ecology and Sociobiology,* 62, 935–43.

MAINGUY, J., COTE, S. D., CARDINAL, E., & HOULE, M. 2008. Mating tactics and mate choice in relation to age and social rank in male mountain goats. *Journal of Mammalogy,* 89, 626–35.

MÁRQUEZ, R. 1995. Female choice in the midwife toads (*Alytes obstetricans* and *A. cisternasii*). *Behaviour,* 132, 151–61.

MÁRQUEZ, R. & VERRELL, P. 1991. The courtship and mating of the Iberian midwife toad *Alytes cisternasii* (Amphibia: Anura: Discoglossidae). *Journal of Zoology,* 225, 125–39.

MARSHALL, D. L. & FOLSOM, M. W. 1991. Mate choice in plants—an anatomical to population perspective. *Annual Review of Ecology, Evolution, and Systematics,* 22, 37–63.

MARTIN, M. D. & MENDELSON, T. C. 2016. Male behaviour predicts trait divergence and the evolution of reproductive isolation in darters (Percidae: Etheostoma). *Animal Behaviour,* 112, 179–86.

MCCLENDON, D. 2018. Crossing boundaries: "Some College," Schools, and Educational Assortative Mating. *Journal of Marriage and Family,* 80, 812–25.

MCLENNAN, D. A. 1995. Male mate choice based upon female nuptial coloration in the brook stickleback, *Culaea inconstans* (Kirtland). *Animal Behaviour,* 50, 213–21.

MEYER, T. B. & UETZ, G. W. 2018. Complex male mate choice in the brush-legged wolf spider *Schizocosa ocreata* (Hentz). *Behavioral Ecology,* 30, 27–38.

MIENO, A. & KARINO, K. 2019. Male mate preference for female coloration in a cyprinid fish, *Puntius titteya. Zoological Science,* 36, 504–10.

MONAGHAN, P., METCALFE, N. B., & HOUSTON, D. C. 1996. Male finches selectively pair with fecund females. *Proceedings of the Royal Society of London—Series B: Biological Sciences,* 263, 1183–6.

MORAN, R. L., ZHOU, M., CATCHEN, J. M., & FULLER, R. C. 2017. Male and female contributions to behavioral isolation in darters as a function of genetic distance and color distance. *Evolution,* 71, 2428–44.

MORSE, P., ZENGER, K. R., MCCORMICK, M. I., MEEKAN, M. G., & HUFFARD, C. L. 2017. Chemical cues correlate with agonistic behaviour and female mate choice in the southern blue-ringed octopus, *Hapalochlaena maculosa* (Hoyle, 1883) (Cephalopoda: Octopodidae). *Journal of Molluscan Studies,* 83, 79–87.

MULLER, M. N., THOMPSON, M. E., & WRANGHAM, R. W. 2006. Male chimpanzees prefer mating with old females. *Current Biology,* 16, 2234–8.

NAGEL, R., KIRSCHBAUM, F., ENGELMANN, J., HOFMANN, V., PAWELZIK, F., & TIEDEMANN, R. 2018. Male-mediated species recognition among African weakly electric fishes. *Royal Society Open Science,* 5, 170443.

NG, S. Y., BHARDWAJ, S. & MONTEIRO, A. 2017. Males become choosier in response to manipulations of female wing ornaments in dry season *Bicyclus anynana* butterflies. *Journal of Insect Science,* 17, 1–6.

NG, T. P., SALTIN, S. H., DAVIES, M. S., JOHANNESSON, K., STAFFORD, R., & WILLIAMS, G. A. 2013. Snails and their trails: the multiple functions of trail-following in gastropods. *Biological Reviews,* 88, 683–700.

NG, T. P. T. & WILLIAMS, G. A. 2014. Size-dependent male mate preference and its association with size-assortative mating in a mangrove snail, *Littoraria ardouiniana. Ethology,* 120, 995–1002.

NICHOLS, H. J., AMOS, W., CANT, M. A., BELL, M. B., & HODGE, S. J. 2010. Top males gain high reproductive success by guarding more successful females in a cooperatively breeding mongoose. *Animal Behaviour,* 80, 649–57.

NIEUWENHUIS, B. P. S. & AANEN, D. K. 2012. Sexual selection in fungi. *Journal of Evolutionary Biology,* 25, 2397–411.

NOBILE, C. & JOHNS, P. 2005. Prelude to a kiss: evidence for mate discrimination in the striped bark scorpion, *Centruroides vittatus. Journal of Insect Behavior,* 18, 405–13.

NUTTALL, D. B. & KEENLEYSIDE, M. H. 1993. Mate choice by the male convict cichlid (*Cichlasoma nigrofasciatum*; Pisces, Cichlidae). *Ethology,* 95, 247–56.

OGDEN, H. J. P., DE BOER, R. A., DEVIGILI, A., REULAND, C., KAHRL, A. F., & FITZPATRICK, J. L. 2019. Male mate choice for large gravid spots in a live-bearing fish. *Behavioral Ecology,* 31, 63–72.

OLSSON, M. 1993. Male preference for large females and assortative mating for body size in the sand lizard (*Lacerta agilis*). *Behavioral Ecology and Sociobiology,* 32, 337–41.

ORBACH, D. N., PACKARD, J. M., & WUERSIG, B. 2014. Mating group size in dusky dolphins (*Lagenorhynchus obscurus*): costs and benefits of scramble competition. *Ethology,* 120, 804–15.

ORRELL, K. S. & JENSSEN, T. A. 2002. Male mate choice by the lizard *Anolis carolinensis*: a preference for novel females. *Animal Behaviour,* 63, 1091–102.

PACK, A. A., HERMAN, L. M., SPITZ, S. S., CRAIG, A. S., HAKALA, S., DEAKOS, M. H., HERMAN, E. Y. K., MILETTE, A. J., CARROLL, E., LEVITT, S., & LOWE, C. 2012. Size-assortative pairing and discrimination of potential mates by humpback whales in the Hawaiian breeding grounds. *Animal Behaviour,* 84, 983–93.

PAL, P., ERLANDSSON, J., & SKOLD, M. 2006. Size-assortative mating and non-reciprocal copulation in a hermaphroditic intertidal limpet: test of the mate availability hypothesis. *Marine Biology,* 148, 1273–82.

PARGA, J. A. 2003. Copulatory plug displacement evidences sperm competition in *Lemur catta. International Journal of Primatology,* 24, 889–99.

PARGA, J. A. 2006. Male mate choice in *Lemur catta. International Journal of Primatology,* 27, 107.

PARGA, J. A. 2010. Male post-ejaculatory mounting in the ring-tailed lemur (*Lemur catta*): a behavior solicited by females? *Ethology,* 116, 832–42.

PASSOS, C., TASSINO, B., REYES, F., & ROSENTHAL, G. G. 2014. Seasonal variation in female mate choice and operational sex ratio in wild populations of an annual fish, *Austrolebias reicherti. PLoS ONE,* 9.

PASSOS, C., VIDAL, N., & D'ANATRO, A. 2019. Male mate choice in the annual fish *Austrolebias reicherti* (Cyprinodontiformes: Rivulidae): when size matters. *Journal of Ethology,* 37, 1–6.

PELABON, C., BORG, A. A., BJELVENMARK, J., FORSGREN, E., BARBER, I., & AMUNDSEN, T. 2003. Do male two-spotted gobies prefer large fecund females? *Behavioral Ecology,* 14, 787–92.

PERETTI, A. V., ACOSTA, L. E., & BENTON, T. G. 1999. Sexual cannibalism in scorpions: fact or fiction? *Biological Journal of the Linnean Society,* 68, 485–96.

PIEROTTI, M. E., MARTÍN-FERNÁNDEZ, J. A., & SEEHAUSEN, O. 2009. Mapping individual variation in male mating preference space: multiple choice in a color polymorphic cichlid fish. *Evolution: International Journal of Organic Evolution,* 63, 2372–88.

POLLO, P., MUNIZ, D. G., & SANTOS, E. S. 2019. Be prudent if it fits you well: male mate choice depends on male size in a golden orb-weaver spider. *Animal Behaviour,* 156, 11–20.

POULIN, R. 1994. Metaanalysis of parasite-induced behavioral-changes. *Animal Behaviour,* 48, 137–46.

PRESTON, B. T., STEVENSON, I. R., PEMBERTON, J. M., COLTMAN, D. W., & WILSON, K. 2005. Male mate choice influences female promiscuity in Soay sheep. *Proceedings of the Royal Society of London—Series B: Biological Sciences,* 272, 365–73.

PROKOP, P. 2019. Male preferences for nuptial gifts and gift weight loss amongst the nursery web spider, *Pisaura mirabilis. Journal of Ethology,* 37, 1–8.

PRYKE, S. R. & GRIFFITH, S. C. 2007. The relative role of male vs. female mate choice in maintaining assortative pairing among discrete colour morphs. *Journal of Evolutionary Biology,* 20, 1512–21.

PYRON, M. 1996. Male orangethroat darters, *Etheostoma spectabile*, do not prefer larger females. *Environmental Biology of Fishes,* 47, 407–10.

RABB, G. B., WOOLPY, J. H., & GINSBURG, B. E. 1967. Social relationships in a group of captive wolves. *American Zoologist*, 7, 305–11.

READING, K. L. & BACKWELL, P. R. 2007. Can beggars be choosers? Male mate choice in a fiddler crab. *Animal Behaviour*, 74, 867–72.

REMEŠ, V. & MATYSIOKOVÁ, B. 2013. More ornamented females produce higher-quality offspring in a socially monogamous bird: an experimental study in the great tit (*Parus major*). *Frontiers in Zoology*, 10, 14.

REYES-RAMÍREZ, A., SANDOVAL-GARCÍA, I. A., ROCHA-ORTEGA, M., & CÓRDOBA-AGUILAR, A. 2020. Mutual mate choice and its benefits for both sexes. *Scientific Reports*, 10, 1–8.

RIECHERT, S. E. & SINGER, F. D. 1995. Investigation of potential male mate choice in a monogamous spider. *Animal Behaviour*, 49, 715–23.

RIESCH, R., MUSCHICK, M., LINDTKE, D., VILLOUTREIX, R., COMEAULT, A. A., FARKAS, T. E., LUCEK, K., HELLEN, E., SORIA-CARRASCO, V., DENNIS, S. R., DE CARVALHO, C. F., SAFRAN, R. J., SANDOVAL, C. P., FEDER, J., GRIES, R., CRESPI, B. J., GRIES, G., GOMPERT, Z., & NOSIL, P. 2017. Transitions between phases of genomic differentiation during stick-insect speciation. *Nature Ecology & Evolution*, 1, 0082.

ROSENTHAL, M. F., GERTLER, M., HAMILTON, A. D., PRASAD, S., & ANDRADE, M. C. 2017. Taxonomic bias in animal behaviour publications. *Animal Behaviour*, 127, 83–9.

ROWLAND, W. J. 1982. Mate choice by male sticklebacks, *Gasterosteus aculeatus*. *Animal Behaviour*, 30, 1093–8.

ROWLAND, W. J. 1989. The effects of body size, aggression and nuptial coloration on competition for territories in male threespine sticklebacks, *Gasterosteus aculeatus*. *Animal Behaviour*, 37, 282–9.

ROWLAND, W. J., BAUBE, C. L., & HORAN, T. T. 1991. Signaling of sexual receptivity by pigmentation pattern in female sticklebacks. *Animal Behaviour*, 42, 243–9.

ROWLAND, W. J. & SEVENSTER, P. 1985. Sign stimuli in the threespine stickleback (*Gasterosteus aculeatus*): a re-examination and extension of some classic experiments. *Behaviour*, 93, 241–57.

SALTIN, S. H., SCHADE, H., & JOHANNESSON, K. 2013. Preference of males for large females causes a partial mating barrier between a large and a small ecotype of *Littorina fabalis* (W. Turton, 1825). *Journal of Molluscan Studies*, 79, 128–32.

SANDERSON, J. L., WANG, J., VITIKAINEN, E. I. K., CANT, M. A., & NICHOLS, H. J. 2015. Banded mongooses avoid inbreeding when mating with members of the same natal group. *Molecular Ecology*, 24, 3738–51.

SANTOS, P. S. C., MICHLER, F. U., & SOMMER, S. 2017. Can MHC-assortative partner choice promote offspring diversity? A new combination of MHC-dependent behaviours among sexes in a highly successful invasive mammal. *Molecular Ecology*, 26, 2392–404.

SARGENT, R. C., GROSS, M. R., & VAN DEN BERGHE, E. P. 1986. Male mate choice in fishes. *Animal Behaviour*, 34, 545–50.

SATO, N., KASUGAI, T., & MUNEHARA, H. 2014. Female pygmy squid cryptically favour small males and fast copulation as observed by removal of spermatangia. *Evolutionary Biology*, 41, 221–8.

SATO, N., YOSHIDA, M., & KASUGAI, T. 2017. Impact of cryptic female choice on insemination success: larger sized and longer copulating male squid ejaculate more, but females influence insemination success by removing spermatangia. *Evolution*, 71, 111–20.

SCHERER, U., KUHNHARDT, M., & SCHUETT, W. 2018. Predictability is attractive: female preference for behaviourally consistent males but no preference for the level of male aggression in a bi-parental cichlid. *PLoS ONE*, 13.

SCHERER, U. & SCHUETT, W. 2018. No male mate choice for female boldness in a bi-parental West African cichlid, the rainbow krib (*Pelvicachromis pulcher*). *Peerj*, 6, e5373.

SCHIESTL, F. P., AYASSE, M., PAULUS, H. F., LÖFSTEDT, C., HANSSON, B. S., IBARRA, F., & FRANCKE, W. 1999. Orchid pollination by sexual swindle. *Nature*, 399, 421.

SCHLAEPFER, M. A., RUNGE, M. C., & SHERMAN, P. W. 2002. Ecological and evolutionary traps. *Trends in Ecology & Evolution*, 17, 474–80.

SCHLUPP, I. 2018. Male mate choice in livebearing fishes: an overview. *Current Zoology*, 64, 393–403.

SCHNELL, A. K., SMITH, C. L., HANLON, R. T., & HARCOURT, R. T. 2015. Female receptivity, mating history, and familiarity influence the mating behavior of cuttlefish. *Behavioral Ecology and Sociobiology*, 69, 283–92.

SCHWAGMEYER, P. & PARKER, G. 1990. Male mate choice as predicted by sperm competition in thirteen-lined ground squirrels. *Nature*, 348, 62.

SETCHELL, J. M. & JEAN WICKINGS, E. 2006. Mate choice in male mandrills (*Mandrillus sphinx*). *Ethology*, 112, 91–9.

SETCHELL, J. M., RICHARDS, S. A., ABBOTT, K. M., & KNAPPE, L. A. 2016. Mate-guarding by male mandrills (*Mandrillus sphinx*) is associated with female MHC genotype. *Behavioral Ecology*, 27, 1756–66.

SHINE, R., PHILLIPS, B., WAYE, H., LEMASTER, M., & MASON, R. T. 2004. Species-isolating mechanisms in a mating system with male mate choice (garter snakes, *Thamnophis* spp.). *Canadian Journal of Zoology*, 82, 1091–8.

SHINE, R., WEBB, J. K., LANE, A., & MASON, R. T. 2006. Flexible mate choice: a male snake's preference for larger females is modified by the sizes of females encountered. *Animal Behaviour*, 71, 203–9.

SKOGSMYR, I. & LANKINEN, A. 2002. Sexual selection: an evolutionary force in plants. *Biological Reviews*, 77, 537–62.

SOUDRY, O., KAIRA, H., PARSA, S., & MENDELSON, T. 2020. Male rainbow darters (*Etheostoma caeruleum*) prefer larger conspecific females. *Behavioural Processes*, 170, 104013.

SQUIRES, Z. E., WONG, B. B. M., NORMAN, M. D., & STUART-FOX, D. 2014. Multiple paternity but no evidence of biased sperm use in female dumpling squid *Euprymna tasmanica*. *Marine Ecology Progress Series*, 511, 93–103.

SQUIRES, Z. E., WONG, B. B. M., NORMAN, M. D., & STUART-FOX, D. 2015. Last male sperm precedence in a polygamous squid. *Biological Journal of the Linnean Society*, 116, 277–87.

STÜMPKE, H. 1975. Bau und Leben der Rhinogradentia. Heidelberg, Germany: Spektrum Akademischer Verlag.

SUMMERS, K. 2014. Sexual conflict and deception in poison frogs. *Current Zoology*, 60, 37–42.

SUMMERS, K. & MCKEON, C. S. 2004. The evolutionary ecology of phytotelmata use in Neotropical poison frogs. *Miscellaneous Publications Museum of Zoology University of Michigan*, 193, 55–73.

SUMMERS, K. & TUMULTY, J. 2014. Parental care, sexual selection, and mating systems in neotropical poison frogs. In MACEDO, R. H. & MACHADO, G. (eds.) *Sexual Selection: Perspectives and Models from the Neotropics*. Amsterdam: Elsevier.

SWIERK, L. & LANGKILDE, T. 2019. Fitness costs of mating with preferred females in a scramble mating system. *Behavioral Ecology*, 30, 658–65.

SWIERK, L., MYERS, A., & LANGKILDE, T. 2013. Male mate preference is influenced by both female behaviour and morphology. *Animal Behaviour*, 85, 1451–7.

SWIERK, L., RIDGWAY, M., & LANGKILDE, T. 2012. Female lizards discriminate between potential reproductive partners using multiple male traits when territory cues are absent. *Behavioral Ecology and Sociobiology*, 66, 1033–43.

SYNNOTT, A. L., FULKERSON, W., & LINDSAY, D. 1981. Sperm output by rams and distribution amongst ewes under conditions of continual mating. *Journal of Reproduction and Fertility*, 61, 355–61.

SZYKMAN, M., VAN HORN, R., ENGH, A., BOYDSTON, E., & HOLEKAMP, K. 2007. Courtship and mating in free-living spotted hyenas. *Behaviour*, 144, 815–46.

THRASHER, P., REYES, E., & KLUG, H. 2015. Parental care and mate choice in the giant water bug *Belostoma lutarium*. *Ethology*, 121, 1018–29.

THÜNKEN, T., BAKKER, T. C. M., & BALDAUF, S. A. 2014. "Armpit effect" in an African cichlid fish: self-referent kin recognition in mating decisions of male *Pelvicachromis taeniatus*. *Behavioral Ecology and Sociobiology*, 68, 99–104.

THÜNKEN, T., MEUTHEN, D., BAKKER, T. C. M., & BALDAUF, S. A. 2012. A sex-specific trade-off between mating preferences for genetic compatibility and body size in a cichlid fish with mutual mate choice. *Proceedings of the Royal Society of London—Series B: Biological Sciences*, 279, 2959–64.

THURMAN, T. J., BRODIE, E., EVANS, E., & MCMILLAN, W. O. 2018. Facultative pupal mating in *Heliconius erato*: implications for mate choice, female preference, and speciation. *Ecology and Evolution*, 8, 1882–9.

TOKARZ, R. R. 1992. Male mating preference for unfamiliar females in the lizard, *Anolis sagrei*. *Animal Behaviour*, 44, 843–9.

TONNABEL, J., DAVID, P., & PANNELL, J. R. 2019. Do metrics of sexual selection conform to Bateman's principles in a wind-pollinated plant? *Proceedings of the Royal Society of London—Series B: Biological Sciences*, 286, 20190532.

TRAVIS, J. 2020. Where is natural history in ecological, evolutionary, and behavioral science? *The American Naturalist*, 196, 1–8.

TUDOR, S. & MORRIS, M. 2009. Variation in male mate preference for female size in the swordtail *Xiphophorus malinche*. *Behaviour*, 146, 727–40.

VÁGI, B., VÉGVÁRI, Z., LIKER, A., FRECKLETON, R. P., & SZÉKELY, T. 2019. Parental care and the evolution of terrestriality in frogs. *Proceedings of the Royal Society of London—Series B: Biological Sciences*, 286, 20182737.

VALLEJOS, J. G., GRAFE, T. U., SAH, H. H. A., & WELLS, K. D. 2017. Calling behavior of males and females of a Bornean frog with male parental care and possible sex-role reversal. *Behavioral Ecology and Sociobiology*, 71, 95.

VAN DEN BERGHE, E. & WARNER, R. 1989. The effects of mating system on male mate choice in a coral reef fish. *Behavioral Ecology and Sociobiology*, 24, 409–15.

VARRICCHIO, D. J., MOORE, J. R., ERICKSON, G. M., NORELL, M. A., JACKSON, F. D., & BORKOWSKI, J. J. 2008. Avian paternal care had dinosaur origin. *Science*, 322, 1826–8.

VEIGA, J. P. 1990. Sexual conflict in the house sparrow: interference between polygynously mated females versus asymmetric male investment. *Behavioral Ecology and Sociobiology*, 27, 345–50.

VELANDO, A., EIROA, J., & DOMÍNGUEZ, J. 2008. Brainless but not clueless: earthworms boost their ejaculates when they detect fecund non-virgin partners. *Proceedings of the Royal Society of London—Series B: Biological Sciences*, 275, 1067–72.

VENNER, S., BERNSTEIN, C., DRAY, S., & BEL-VENNER, M. C. 2010. Make love not war: when should less competitive males choose low-quality but defendable females? *American Naturalist,* 175, 650–61.

VERRELL, P. A. 1985. Male mate choice for large, fecund females in the red-spotted newt, *Notophthalmus viridescens*: how is size assessed? *Herpetologica,* 1985, 382–6.

VERRELL, P. A. 1986. Male discrimination of larger, more fecund females in the smooth newt, *Triturus vulgaris. Journal of Herpetology,* 1986, 416–22.

VERRELL, P. A. 1989. Male mate choice for fecund females in a plethodontid salamander. *Animal Behaviour,* 38, 1086–8.

VERRELL, P. A. 1994. Males may choose larger females as mates in the salamander *Desmognathus fuscus. Animal Behaviour,* 47, 1465–7.

VERRELL, P. A. 1995. Males choose larger females as mates in the salamander *Desmognathus santeetlah. Ethology,* 99, 162–71.

VERRELL, P. A. & BROWN, L. E. 1993. competition among females for mates in a species with male parental care, the midwife toad *Alytes obstetricans. Ethology,* 93, 247–57.

WANG, D., KEMPENAERS, N., KEMPENAERS, B., & FORSTMEIER, W. 2017a. Male zebra finches have limited ability to identify high-fecundity females. *Behavioral Ecology,* 28, 784–92.

WANG, D. P., FORSTMEIER, W., & KEMPENAERS, B. 2017b. No mutual mate choice for quality in zebra finches: time to question a widely held assumption. *Evolution,* 71, 2661–76.

WARD, P. I. 1988. Sexual selection, natural selection, and body size in *Gammarus pulex* (Amphipoda). *The American Naturalist,* 131, 348–59.

WEGENER, B. J., STUART-FOX, D. M., NORMAN, M. D., & WONG, B. B. 2013. Strategic male mate choice minimizes ejaculate consumption. *Behavioral Ecology,* 24, 668–71.

WEISS, S. L. & DUBIN, M. 2018. Male mate choice as differential investment in contest competition is affected by female ornament expression. *Current Zoology,* 64, 335–44.

WELLS, K. D. 1977. The social behaviour of anuran amphibians. *Animal Behaviour,* 25, 666–93.

WEN, Y. H. 1993. Sexual dimorphism and mate choice in *Hyalella azteca* (Amphipoda). *American Midland Naturalist,* 129, 153–60.

WERNER, N. Y. & LOTEM, A. 2003. Choosy males in a haplochromine cichlid: first experimental evidence for male mate choice in a lekking species. *Animal Behaviour,* 66, 293–8.

WERNER, N. Y. & LOTEM, A. 2006. Experimental evidence for male sequential mate preference in a lekking species. *Ethology,* 112, 657–63.

WEST, R. J. D. & KODRIC-BROWN, A. 2015. Mate choice by both sexes maintains reproductive isolation in a species flock of pupfish (*Cyprinodon* spp.) in the Bahamas. *Ethology,* 121, 793–800.

WHITING, M. J. & BATEMAN, P. W. 1999. Male preference for large females in the lizard *Platysaurus broadleyi. Journal of Herpetology,* 33, 309–12.

WILLSON, M. F. 1994. Sexual selection in plants: perspective and overview. *The American Naturalist,* 144, S13–39.

WILLSON, M. F. & BURLEY, N. 1983. *Mate Choice in Plants: Tactics, Mechanisms, and Consequences.* Princeton, NJ: Princeton University Press.

WILSON, E. O. 1975. *Sociobiology. The Modern Synthesis.* Cambridge, MA: Belknap.

WINGFIELD, J. C. 2017. The challenge hypothesis: where it began and relevance to humans. *Hormones and Behavior,* 92, 9–12.

WITTE, K. & CURIO, E. 1999. Sexes of a monomorphic species differ in preference for mates with a novel trait. *Behavioral Ecology,* 10, 15–21.

WITTE, K. & SAWKA, N. 2003. Sexual imprinting on a novel trait in the dimorphic zebra finch: sexes differ. *Animal Behaviour,* 65, 195–203.

WONG, B. B. & JENNIONS, M. D. 2003. Costs influence male mate choice in a freshwater fish. *Proceedings of the Royal Society of London—Series B: Biological Sciences,* 270, S36–8.

WRIGHT, D. S., PIEROTTI, M. E., RUNDLE, H. D., & MCKINNON, J. S. 2015. Conspicuous female ornamentation and tests of male mate preference in threespine sticklebacks (*Gasterosteus aculeatus*). *PLoS ONE,* 10, e0120723.

WRONSKI, T., BIERBACH, D., CZUPALLA, L.-M., LERP, H., ZIEGE, M., CUNNINGHAM, P. L., & PLATH, M. 2012. Rival presence leads to reversible changes in male mate choice of a desert dwelling ungulate. *Behavioral Ecology,* 23, 551–8.

WYMANN, M. N. & WHITING, M. J. 2003. Male mate preference for large size overrides species recognition in allopatric flat lizards (*Platysaurus broadleyi*). *Acta Ethologica,* 6, 19–22.

WYNN, S. E. & PRICE, T. 1993. Male and female choice in zebra finches. *The Auk,* 1993, 635–8.

YIĞIT, A., SEVGILI, H., & ÖZDEMIR, H. 2019. Male bush-crickets in female-biased environment allocate fewer sperm per ejaculation. *Entomological News,* 128, 393–403.

YONG, L., LEE, B., & MCKINNON, J. S. 2018. Variation in female aggression in 2 three-spined stickleback populations with female throat and spine coloration *Current Biology,* 64, 345–50.

CHAPTER 4

What Males Choose: Differences in Female Quality

4.1 Brief outline of the chapter

In female choice, differences in male quality are very important. Males display to females to provide information often via costly ornaments. Females also differ in quality, but what they display to males is less clear. Also, how males evaluate differences in female quality is not well understood. From the literature on male mate choice one might conclude that female fecundity is the most important feature a female can display to a choosy male. But I argue that there must be many more features of females that are important in male mate choice, maybe even indirect benefits.

4.2 Introduction

When females evaluate partners as potential mates, many different traits of males have been identified as meaningful in that decision process. However, based on our review of examples from Chapter 3, males seem to pay attention most often to one aspect, female fecundity with body size as the actual trait evaluated. While this may be true, I find it more likely that more research will reveal that males use relevant information about females just like females use all kinds of information about males.

Female choice seems to happen under at least one of three scenarios: (i) there is a direct benefit to the female, (ii) there is an indirect (genetic) benefit to the females, or (iii) they select a partner who is a good genetic or social fit for them. It seems reasonable to assume that males evaluate their partners along the same lines. An important question in this context is whether male mate choice can drive the evolution of female ornaments. After all, if females differ in quality, they might advertise these differences. This will be covered in Chapter 7.

4.3 Direct benefits

Unsurprisingly, there is strong support for male choice for female quality. Males—just like females—surely derive benefits from mating with higher-quality partners as predicted by theory. The definition of quality is not always clear, but for the purpose of this book it could be any trait that positively influences male fitness. Females are thought to benefit from choosiness in three different ways: direct benefits, indirect benefits, and compatibility. In principle this should also apply to male mate choice. The evidence for direct benefits is quite strong, while indirect benefits and compatibility have barely been explored.

The original idea that larger females have a fecundity advantage goes back to Darwin's original thinking about sexual selection (Darwin, 1871). Although almost universally accepted, the concept has been criticized as too broad by Shine (1988). He pointed out that one needs to investigate lifetime reproductive success, which may be less influenced by body size than individual clutches. Also, reproductive output by females is clearly not only influenced by female size but vice versa, female size may influence many offspring related traits. Offspring of larger females may be larger, for example, but not more numerous. They may, of course, be of better "quality." Research on cave mollies, a population of the Atlantic molly (*Poecilia mexicana*), for example, has shown that females from different populations can have significantly different numbers of offspring

Male Choice, Female Competition, and Female Ornaments in Sexual Selection. Ingo Schlupp, Oxford University Press (2021). © Ingo Schlupp (2021).
DOI: 10.1093/oso/9780198818946.003.0004

at the same female size (Riesch et al., 2009, 2010). But within a population, the relationship between female size and offspring number is still positive, although the slopes are different (Riesch et al., 2016). In other words, male choice for larger females would still be beneficial, but the fitness return might differ by population. Male preferences for female size in these populations did not reflect this, however. Males preferred larger females, but did not respond to the differences in fecundity (Arriaga and Schlupp, 2013). Perhaps, because the correlation between female size and fecundity is generally so strong, this is one of the most obvious explanations for why males should prefer certain females. In other words, males may benefit from preferring larger females in more ways than just a fecundity benefit. But clearly, there must be other aspects of female quality that males might pay attention to.

4.3.1 Direct benefits: fecundity

A very large number of studies found male preferences for female size. This is usually interpreted as a fecundity benefit because female size is almost always positively correlated with fecundity. However, there are other interpretations for this pattern as well. First, a preference for larger may be trivial, as bigger is generally better (Barry and Kokko, 2010). Barry and Kokko argue that finding male preferences in simultaneous mate choice trials could be almost trivial because a larger female is practically always more fecund and hence more valuable for a male, but a simultaneous choice situation may not be realistic under natural conditions. This calls for more studies of preference functions (Wagner, 1998; Schlupp, 2018). Furthermore, almost all studies of male preferences only state that there is a preference for larger females, but actual fitness benefits are not documented. Also, few studies examine the mating preferences for both males and females in one mating system. This would be very useful to better understand the potential role of genetic correlations. It would be of great interest to study species where, for example, females have a preference for larger males, but males do not show a preference for larger females. I will note again that the default explanation for preferences for larger partners differs based on sex: male preferences are

thought to be governed by direct benefits, whereas female preferences are often attributed to indirect benefits (Schlupp, 2018).

Furthermore, the expression of male preferences might be a pleiotropic effect of the evolved preferences found in females. This is an interesting argument because it could lead to the situation described above: the female preference for larger males is thought to yield indirect benefits for the females, whereas a preference for more fecund females would result in a direct benefit. A shared genetic preference for "large" may also have been selected for in males and is expressed in females as a pleiotropic effect. If indeed the preference for "large" represents a pleiotropic effect, it should be adaptive in one of the sexes, while it does not have to be in the other sex. This can be tested directly by studying the fitness consequences of mate choice in both sexes of the same species.

The observed preferences could also be byproducts of existing sensory biases that generally lead animals to prefer larger alternatives (Ryan and Keddy-Hector, 1992). This has been hypothesized for females and might also apply to males (Chapter 2).

Finally, we should also consider that relatively few studies have investigated male preferences for other traits than a fecundity benefit because it is such an obvious explanation for male preferences. This may lead to other adaptive explanations being overlooked or ignored in the formation of hypotheses and the design of experiments. A result of this could be a biased literature. In addition, there are likely unpublished studies of male preferences that were inconclusive, or did not find male preferences for female size, potentially adding to any publication bias. Most importantly, fecundity is only one way—albeit an obvious one—females can differ in quality. They can also differ in several other ways, for example in age, health status, degree of experienced sperm competition, mating status, or general condition. Nonetheless, of course, the relationship of size and fecundity is strong. In red king crab, *Paralithodes camtschaticus*, for example, a strong relationship was found over several years (Swiney et al., 2012) (Figure 4.1).

In addition to better understanding the ultimate consequences of male mate choice, we also need to

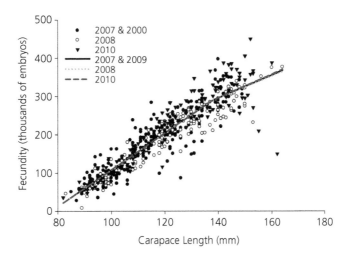

Figure 4.1 Size–fecundity relationship in the red king crab, *Paralithodes camtschaticus*, over several years (Swiney et al., 2012). Original artwork: Katherine Swiney. First appeared in SWINEY, K. M., LONG, W. C., ECKERT, G. L., & KRUSE, G. H. 2012. Red king crab, *Paralithodes camtschaticus*, size–fecundity relationship, and interannual and seasonal variability in fecundity. *Journal of Shellfish Research*, 31, 925–33. Reprinted by kind permission.

look into the mechanisms and discover how males assess differences in female quality. In field crickets (*Teleogryllus oceanicus*) for example, female fecundity is correlated with certain chemicals from the cocktail of cuticular hydrocarbons many insects display (Berson and Simmons, 2019). Males and females differ in their cuticular hydrocarbons and males show preferences for more fecund females, using chemical information as information source (Thomas and Simmons, 2010).

4.3.2 Direct benefits: other than fecundity

Future reproductive value

While in some mating systems, mating may be a one-time decision for the mating partners, in others it is not. If males mate multiple times, they face a tradeoff between current and future investment, leading to more complex decision making processes. For example, this is the case in monogamous mating systems.

Timing

Many females mate multiple times with different males and consequently sperm of several males may be present in the female's genital tract at the same time—provided they have internal fertilization, leading to both sperm competition (Parker, 1970; Birkhead and Hunter, 1990) and cryptic female choice (Firman et al., 2017). There are several explanations as to why such multiple matings may be adaptive for females. Also, and maybe more importantly, in many systems the order in which males mate with females does influence the likelihood of successfully having offspring with a female. In some cases, the last male has a higher probability to sire offspring, in others it is the first male. In those cases, the order in which males inseminate a female becomes potentially important and males should choose based on information on which other males have mated with the female before or will likely mate with her after. Often such information will be transmitted socially (Chapter 2). Females, of course may benefit from concealing previous matings and from misleading about future matings. Furthermore, from a male perspective it might be adaptive to manipulate the information other males have about a mating and try to deceive competing males.

Differences in mating status: unmated females

In many species, males prefer to mate with females that have had no previous matings, sometimes called virgin females. In species that mate only once, the benefit is obvious as the successful male will have reproductive success, whereas other males will not. This might be the case in pupal mating in some butterflies where males guard females prior to eclosion and mate with them as soon as they emerge from their cocoon. An example for this can be found in some species of the tropical genus *Heliconius* (Mendoza-Cuenca and Macias-Ordonez, 2010). Single matings can be because of male manipulation, which forces females into *just* single matings. This sexual conflict can play out with several mechanisms ranging, for example, from mate guarding via extended mating duration to mating plugs that evolved to mechanically prevent other males from mating with a female. Males may even add a chemical that repels other males from mating with a female (Andersson et al., 2000). From a male point of view mating with a virgin is especially beneficial if this is the only mating this female ever experiences, in particular, as this also avoids any sperm competition. Even in species that mate multiple times, the first sperm may be stored and used again for more fertilizations. Males should typically allocate relatively small ejaculates to virgin females, but not under all circumstances: if female mating rate is high, males should invest larger ejaculates (Engqvist and Reinhold, 2006).

In other species a preference for virgins is less easy to understand. In guppies (*Poecilia reticulata*), for example, males prefer virgin females, but need close-range information to determine which female is virgin (Guevara-Fiore et al., 2009). This is somewhat puzzling because female guppies keep growing—actually indefinitely—and older females are therefore more fecund. However, since the females store sperm, mating with larger females also means stronger sperm competition. Consequently, paternity assurance and avoidance of sperm competition seem to be important, but on the other hand sperm competition is ubiquitous and the intensity of it seems to be difficult to accurately assess for males. Only old females that are showing signs of senescence should be rejected by males.

In a parasitoid wasp, *Spalangia endius*, virgin females were preferred by both virgin and nonvirgin males (King et al., 2005). In this species mated females rarely mate again, providing a clear benefit

for mating with virgin females. Also, in mealworm beetles (*Tenebrio molitor*) males prefer virgin females (Carazo et al., 2004). The same is true for some spiders (Riechert and Singer, 1995; Gaskett, 2007). Females may potentially benefit from advertising their mating status. They would benefit the most if they could manipulate male perception of their status. This might be the case in a spider (*Schizocosa malitiosa*), where male mate choice was hypothesized but not found (Baruffaldi and Costa, 2014).

In humans (*Homo sapiens*), men's preference for virgin women is well documented (Buss, 2007; Buss and Schmitt, 2019) and is likely influenced by paternity assurance. As always with humans, however, strong cultural influences on this preference are likely, too.

In addition to being virgin or not, females can differ in their mating status based on a reproductive cycle. Independent of the length of the cycle and how it is regulated, it will lead to a situation where females may be inseminated with resulting paternity or not. The time window during which an egg can be successfully fertilized is often rather short, whereas survival of sperm in the female genital tract is often much longer. In humans, for example, eggs live for less than a day, whereas sperm lasts for several days. Humans—like many other species—seem to be able to adjust ejaculate size, and increase the number of sperm present after being physically separated from a partner (Goetz et al., 2007). Interestingly, in horses (*Equus caballos*), stallions provided ejaculates with more motile sperm and higher volume when they mated with a mare that was in estrus (Jeannerat et al., 2018).

Ejaculate adjustment as cryptic male choice

Just like female choice, male choice does not have to be overt. Instead it can be hidden from the female and serve primarily the male. Cryptic male choice can be exercised for example by allocating sperm and ejaculates differentially based on perceived differences in female quality. Allocating sperm strategically may also protect males from becoming sperm depleted. In guppies, males change their behavior after being stripped of their sperm (which is the equivalent to maximal sperm depletion) and seeking fewer mating opportunities and decreasing their choosiness (P. Guevara-Fiore, pers. comm., September 18, 2020). This whole line of argument assumes that sperm and ejaculate production are actually not as cheap as traditionally assumed. Sperm depletion has been identified as a significant problem for males, and especially attractive males are at risk. Such high-quality males should consequently be especially under selection to allocate sperm. Males have been known to modify their ejaculates adaptively relative to sperm competition (Birkhead et al., 2008). This selects for males to adjust sperm numbers and ejaculate sizes, as well as for specialized structures and behaviors to remove sperm of competitors, but can also favor male choosiness. There is a large literature on sperm competition from a sexual conflict point of view (Birkhead et al., 2008), showing that males have evolved a number of traits that can be understood in light of sperm competition including mate guarding, copulation length, and the release of components in the ejaculate that are harmful to females. Ultimately, of course, it is the interaction of cryptic female choice and male ejaculate adjustment that determines who is reproducing successfully, as for example shown in *Drosophila* or guppies (*Poecilia reticulata*) (Cardozo et al., 2020).

Faced with sperm competition, males may not only increase the number of sperm they use, but also evolve behaviors that avoid sperm competition altogether, or deceive other males to mate with less desirable females (Plath et al., 2008). Furthermore, males can show adaptive and anticipatory sperm priming, a mechanism that adjusts sperm production based on the social environment and anticipated mating opportunities. This was shown in guppies, where males that were in contact with females primed more sperm and faster sperm (Cardozo et al., 2020). A similar mechanism was also reported for sailfin mollies (*Poecilia latipinna*), which primed more sperm when exposed to females of their own species, as compared with males that were exposed to females of the sperm parasitic Amazon molly (*Poecilia formosa*) (Aspbury and Gabor, 2004a, 2004b; Robinson et al., 2008).

Clearly, males can express preferences by altering or adjusting the ejaculate they provide for females (Wedell et al., 2002). Such alterations can be viewed either as an adaptive response to the probability of experiencing some form of sperm competition, or as a direct response to perceived female quality. Mostly, however, it seems that adjustments to ejaculates or spermatophores are viewed as adaptations in the context of sperm competition (Larsdotter-Mellström and Wiklund, 2009). Adaptive adjustments can be either an increased number of sperm per ejaculate, or a modified composition. The latter is, for example, found in an Australian field cricket, *Teleogryllus oceanicus*, where males provide a spermatophore with more viable sperm to females that had previously mated with another male (Thomas and Simmons, 2007). In another species of cricket (*Isophya rizeensis*), a bush cricket from Turkey, males were found to invest less in spermatophores when the operational sex ratio (OSR) was more female biased (Yiğit et al., 2019).

The situation in the sexual/asexual mating complex of Amazon mollies (*P. formosa*) is relevant to this question. Amazon mollies are sexual parasites that need to mate with males of a sexual species, mainly *P. latipinna* and *P. mexicana* (Schlupp and Riesch, 2011) but the male DNA is not incorporated into the genome of the offspring. Consequently, males are confronted with a choice between females of their own species or the sexual parasite and should avoid matings with the Amazon molly. However, if they do mate with Amazon molly females, they might reduce their cost by providing fewer sperm to the sexual parasite. This was indeed found in both an experimental (Schlupp and Plath, 2005) and a field study (Riesch et al., 2008). This seems to be the only example of adaptive sperm allocation in heterospecific matings. Interestingly, the males can gain a small benefit from mating with Amazon mollies via mate copying: mating with a heterospecific female makes them more attractive to conspecific females. But the benefit comes from being seen to mate with a heterospecific female and, consequently, the males could avoid the cost of actually transferring sperm by not doing so, or by transferring fewer sperm. Indeed, males prime more sperm when presented with a conspecific female as compared with a heterospecific Amazon molly female (Aspbury and Gabor, 2004a). This clearly seems like an adaptive response to protect the male's sperm reserves. A similar response is found relative to female size:

more sperm is primed in the presence of larger conspecific females (Aspbury and Gabor, 2004b) and by smaller males, which may be at the highest risk of sperm competition.

Another example connects male age with adjustment of sperm. In house sparrows, *Passer domesticus*, older males deliver much larger ejaculates than younger males, possibly as a response to competition (Girndt et al., 2019).

Sperm competition

The biology of sperm has been an important topic in evolutionary biology for several centuries (Birkhead and Montgomerie, 2008). Sperm competition can occur when sperm from at least two males are present and competing for fertilizations is both very common and important (Parker, 1970, 1984). If sperm from more than one male is present to fertilize an egg, males should evolve adaptations that favor their own sperm over the sperm of other males. Such adaptations range from removing sperm of other males before providing their sperm, to ejaculates and individual sperm competing—often inside a female—to males attempting to limit the ability of other males to provide sperm. Males compete with each other, but females also evolved to cryptically choose among the sperm from multiple males, setting the stage for sexual conflict (Eberhard, 1996; Firman et al., 2017), including in humans where eggs seem to be able to exercise independent choice using chemical attractants (Fitzpatrick et al., 2020). On the side of conflict, however, for example, in *Drosophila melanogaster*, males may use harmful substances as part of their ejaculate thereby causing a fitness cost to the females to ascertain that other males have a competitive disadvantage. The content of the ejaculate beyond sperm is a wide field for sexual conflict, and is well documented in insects (Sirot et al., 2014; Sirot, 2019). Sperm competition is more easily studied in organisms with internal fertilization or discrete sperm packages but does also occur in organisms with external fertilization. There are two important aspects that intertwine sperm competition with male mate choice. First, males may experience sperm depletion, which should cause them to allocate sperm in the most beneficial way they can, and second, males may experience different degrees of sperm competition and choose females where their sperm is most likely to obtain paternity.

Although sperm or ejaculate production is undoubtedly energetically less expensive than the production of eggs, this does not mean that sperm or ejaculates are in unlimited supply (Dewsbury, 1982; Wedell et al., 2002). The cost of sperm production was first recognized by Dewsbury (1982), and has since been documented for many species. For example, if males become highly sought after because they are of high quality (or just popular), they may become sperm limited and thus benefit from being choosy. This choosiness can be expressed as a certain timing, for example trying to be the last male mating with a female in a species with last male precedence, but it can also mean to forgo matings with low-quality females.

Differences in sperm competition risk

Depending on the mating system and on information males have about any given female males should try to maximize the probability of their sperm fathering any offspring with that female. They may increase the number of sperm transmitted, for example. A potential cost to that may be a higher risk of sperm depletion. Or they may not mate with females where the cost of mating is not outweighed by a sufficient benefit. One male strategy to account for sperm competition is found in cabbage white butterflies (*Pieris rapae*), where males use larger ejaculates with more sperm to inseminate females that have previously been mated (Wedell and Cook, 1999). A similar pattern has been found in several species of damselflies, where males mate much longer with females that have previously mated (Uhia and Rivera, 2005). In many insects adaptive allocation of ejaculates seems especially acute because they eclose with all the sperm they can spend in their lifetime, as opposed to the many species that can replenish their sperm supplies throughout their lifetime. Males may also try to manipulate the situation in their favor using deception. This had been suggested for males of a fish, the Atlantic molly (*Poecilia mexicana*). Here males were thought to feign a preference for a less valuable female mating partner to mislead other males into mating with her (Plath and Schlupp, 2008; Plath et al., 2008).

Experience, age, and performance

Not all females are equally good at raising offspring. Thus, males can benefit from choosing mothers that can and will provide for the offspring they sire with the female. Experience often comes with age, which means that older females may be better at raising offspring. This argument is interesting for at least two reasons. First, in some organisms there is a correlation between age and size. This could mean that in some of the examples where males choose larger females, they automatically choose older and more experienced females. Second, this goes against the prediction that virgin females should be preferred.

In female choice male age is often used as an indicator. Often younger males are preferred, but this preference may be confounded with male mating history (Wedell and Ritchie, 2004). In the bush cricket *Ephippiger ephippiger* the quality of spermatophores declines with male age, but this reflects male decline rather than actual age as older virgin males are able to produce high-quality spermatophores. In male choice, age and mating history are also confounded. In another cricket, *Acanthoplus discoidalis*, males also produce large spermatophores, and—likely because of that—are selected to forgo matings with nonvirgin females, but they do not tailor their spermatophores (Bateman and Ferguson, 2004).

The complexity of male choice is highlighted by a study on guppies (*Poecilia reticulata*) (Guevara-Fiore and Endler, 2018): male mating history modified their preference for female size and the mating tactic used (courtship or sneaky mating attempts). In this case males with prior mating success had an increased preference for larger females.

In a different context, older females are considered to provide more reliable information in mate copying. Female guppies (*P. reticulata*) were more likely to copy the mate choice of larger (and older) females (Amlacher and Dugatkin, 2005). Presumably, older females have more experience and provide better information. This could predict male preferences for more experienced and older females as partners. The optimal age would likely depend on the individual experience a female has. If very young females are less good as mates and older females, too, this predicts an inverted U-shape for a resulting preference function.

Age also plays a major role in human mate choice. Human men tend to prefer younger women as partners, whereas women tend to prefer somewhat older men (Conroy-Beam and Buss, 2019). But age is not the only factor that is important to men (Buss and Schmitt, 2019). In women, self-perceived mate value and body mass index (BMI) are important and also implicated in female competition (Fernandez et al., 2014). Preferences for older men can create tradeoffs. They may be better and more experienced providers, but bodily functions decay with age, including the quality of sperm (Johnson et al., 2015). For men, the peak "market value" in Western societies is around 30 years of age. At that time, they should also be most selective as to whom they mate with. Interestingly, this seems to also be the case in an insect. In a hide beetle, *Dermestes maculatus*, middle-aged males produce the largest number of offspring (Jones and Elgar, 2004).

Choosers can also respond to subtle differences in quality of displays. What is meant here are for example small differences in the execution of motor patterns. These differences may not be related to ornament size or anything else that is easily measured. Because performance is difficult to capture in easy and quantitative measurements, it is not often studied. In general, though, it seems that females like better motor performance (Byers et al., 2010). In humans, for example, kissing is an important part of pair bonding and foreplay, especially for women (Hughes et al., 2007). They prefer a "good kisser" (Meston and Buss, 2009), but this trait is difficult to quantify.

In other animals, too, females can differ in performance, but we know little about if and how males evaluate female performance. One example could be female performance in aggressive interactions (see Chapter 8), or in synchronized behaviors with their male partner. Such synchronized behaviors are found in duetting (Hall, 2009). This behavior is found in some birds, but also insects (LaRue et al., 2015) and is not limited to acoustic signals (Ręk and Magrath, 2016). In some species, males and females sing in a highly synchronized fashion, but the function of this behavior is not entirely clear (Hall, 2004). One hypothesis implicates territorial defense as the function (Seibt and Wickler, 1977), but another potential function for this behavior might be mutual mate assessment. It might also lead to increased stability of partnerships.

Parasites and diseases

Parasites are immensely important in evolution. They are thought to play a strong role in the evolution of recombination via the "Red Queen hypothesis" (Van Valen, 1973; Lively, 2010; Neiman and Koskella, 2010). Support for this idea is strong, but not universal (Tobler and Schlupp, 2008; Kotusz et al., 2014). Parasites also directly affect mate choice. Avoiding parasitized and sick mating partners is well established in female choice (Hamilton and Zuk, 1982; Zuk, 1992). It seems intuitive for females and males to avoid partners that are visibly sick, but as always, the situation is not so simple. For example, if a potential mate does not show symptoms of a disease, are they good mates because they have already overcome the disease (potentially showing promising immune genes, which provides an intersection with major histocompatibility complex-based preferences; Wegner et al., 2003; Milinski, 2006, Wegner et al., 2003), or are they bad mates that have just been lucky so far and should be avoided? In general, however, predicting that parasitized or sick individuals should be avoided seems reasonable. For sexually transmitted diseases, however, one could argue that strong choice for uninfected partners will likely select for strains of a disease that are more cryptic (Knell and Webberley, 2004). In this fascinating field of complex interactions, mate choice decisions may be equally complex. A review found that most often nonparasitized mating partners are preferred (Beltran-Bech and Richard, 2014). Most of the studies reviewed were on female choice, but the authors list nine studies on male mate choice, where either healthy males were choosing between an infected or uninfected female as in a study on the mosquitofish, *Gambusia affinis* (Deaton, 2009). There are also effects of being infected: In one study males of the corn earworm butterfly, *Helicoverpa zea*, were artificially infected with a virus, and were slower in approaching healthy females (Burand and Tan, 2006). Males of the prairie vole, *Microtos ochrogaster*, with a simulated infection showed no preference for specific females (Bilbo et al., 1999). In a review paper Barber and colleagues noted that most studies

documenting a preference do not estimate the fitness consequences of choice (Barber et al., 2000). This is especially important because we know that parasites often manipulate their hosts, so that a behavior may reflect an adaptation for the parasite, not the host (Moore, 2002). For mate choice this means that parasites may not only alter the phenotype and behavior of infected individuals, but also the decision making of the chooser (Barber et al., 2000). For example, in guppies (*P. reticulata*), females infected by a parasite, *Gyrodactylus turnbulli*, are less choosy than uninfected females (López, 1999). Female guppies in general are avoiding parasitized males (Kennedy, 1987). The same is found in pipefish males, where infected males are less choosy (Mazzi, 2004). The complexity of this is revealed in a study on a small amphipod (*Gammarus insensibilis*), where male mate choice is driven by the time females need to molt. Yet, they can distinguish between parasitized and unparasitized females. On top of that, male mate choice is slower when the male itself is infected (Thomas et al., 1996). While we have many examples of female choice under the threat of parasites, male mate choice under the threat of parasites is poorly understood (Knell and Webberley, 2004). In humans it seems that in cultures with more prevalent pathogens, mate choice is more pronounced and more emphasis is given to physical appearance (Gangestad and Buss, 1993).

In general, investigating parasites and mate choice is a good approach as parasites often have a relatively clear fitness cost (Heins et al., 2004; Heins and Baker, 2008). This is especially important for male mate choice, as parasites have been found to reduce fecundity in females, thus diminishing the most important benefit males seem to be seeking.

Condition

The vast majority of females can and will reproduce. For example, lifelong absence of reproduction only occurs in about 5% of all insect populations (Rhainds, 2019). However, females can differ in many other traits, including condition. One simple argument for male mate choice based on condition would be that males choose females that appear to be able to have offspring and provide for them. This is related to fecundity, of course. The key point here is that condition is more ephemeral than other traits. Some female fishes for example prefer males that look well-fed over hungry-looking males (Plath et al., 2005). As with parasite-free mating partners, condition may or may not be a good indicator for fitness, but generally, it seems—again—good to choose a mating partner in good condition. Of course, also as in parasites, the chooser's own condition can influence the choice. This was studied in field crickets, *Teleogryllus commodus*. Females on a higher-quality diet were more sexually responsive as compared with females on a poorer diet (Hunt et al., 2005). Wolf spider females (*Schizocosa ocreata* and *Schizocosa rovneri*) did not show strong effects on mate choice based on the diet they were raised on, but high-food females mated more often with high-food males (Hebets et al., 2008). These studies point toward a potentially very interesting link between epigenetics and mate choice: diet is often thought to induce maternal (and paternal effects) that persist over multiple generations, including in humans (Veenendaal et al., 2013). Male condition is also a factor in female choice in sticklebacks (*Gasterosteus aculeatus*) (Bakker et al., 1999) (Figure 4.2).

In humans (*H. sapiens*), men's preferences for a certain waist-to-hip ratio in women (Singh, 1993) could be viewed as condition dependent, as deviations in both directions could be indicative of lower reproductive fitness. Changes in waist-to-hip ratio do occur over time, but the timeframe may be too long for this to be called based on "condition." Waist-to-hip ratio-based preferences may be intertwined with preferences for healthy- or unhealthy-looking BMIs, but it seems the same general logic applies (Tovée et al., 1999).

Mating with the right species

In general, mating with members of another species is considered to be costly. Male mate choice is therefore at least partly also trying to identify the best species to mate with (Gaskett, 2007). Some earlier work on male mate choice was focusing on this question. A few examples have been summarized by me (Schlupp, 2018). Generally, mating with another species is viewed as costly, especially for females. Consequently, strong selection is thought to favor reproductive barriers. There are, however, some cases where heterospecific matings can be favored. In some swordtails, *Xiphophorus*, a group of fishes from Mexico, females prefer males from a different species (Ryan and Wagner, 1987). This is attributed to sensory bias. In spadefoot toads (*Spea bombifrons* and *Spea multiplicate*), hybridization is adaptive under specific circumstances and for some females. Consequently, the females alter their preference from conspecific to heterospecific (Pfennig, 2007). These are rare phenomena for females, and examples for males showing similar behavior would be interesting.

4.4 Indirect benefits

Studies that provide good support for indirect benefits in female choice are not abundant (Chapter 1 and 2), and this concept has not really been applied to male mate choice yet, although the argument would essentially be the same for females and males. This is an area where future research would be quite fruitful. Many questions are open here. For example, is there an equivalent to the "sexy son hypothesis," maybe the "sexy daughter hypothesis"? How could indirect benefits work in male mate choice and, if they exist, how can we detect them?

4.5 Compatibility

Although somewhat controversial, we do have a growing list of examples of MHC-based female mate choice (Milinski, 2006). If males choose, finding a fitting partner should also be beneficial for

Figure 4.2 Courting three-spined stickleback (*Gasterosteus aculeatus*) pair (picture credit: Manfred Milinski and Theo C.M. Bakker).

males. Indeed, there are a few examples of male mate choice for matching MHC in females (Wedekind, 2018). In humans the data are suggesting that men also recognize MHC and prefer dissimilar partners (Wedekind and Füri, 1997).

Social compatibility might lead to assortative mating (Chapter 2). In some species partners form long-term partnerships that allow pairs to raise offspring more effectively. This is found in humans, but also in geese. In Canada geese (*Branta canadensis*) for example, breeding partners were found to have similar intensities of nest defense behavior (Clermont et al., 2019).

In addition to picking a mate based on MHC or social compatibility, it can also be beneficial to avoid mating with partners that confer genes that would lead to reduced fitness. These are often selfish elements that can alter the sex ratio or even be lethal to some of the offspring. One example for this is the *t*-element in mice, which causes *t/t* homozygote offspring to die in utero (Manser et al., 2015, 2017). This reduced fitness is detrimental to mothers, but also to fathers. There is evidence that females avoid mating with males that carry the *t*-haplotype, but it would be worthwhile to investigate the male side of this, too.

4.6 Postpairing male mate choice: choosiness when having an affair

Females and males of many species engage in extra-pair copulations. For females one widely accepted hypothesis is that females can gain direct or indirect benefits from this that are unavailable to them otherwise. There are many examples of females mating with "attractive" males supposedly for indirect benefits, while retaining direct benefits from a less "attractive" male. One example is provided by Kempenaers and colleagues in their studies on blue tits (*Parus caeruleus*, now *Cyanistes caeruleus*), which shows female preferences for high-quality males (Kempenaers et al., 1992, 1997). Extra-pair paternity is widespread in monogamous birds and evolutionarily important as—on average—more than 10% of all offspring are the result of extra-pair copulations (Griffith et al., 2008). At least in birds this seems to be largely driven by female choice, including cryptic female choice (Girndt et al., 2018). Interestingly, in this study, males sometimes refuse to mate with

females (A. Girndt, pers. comm., February 14, 2019), presumably because they are sperm limited (Birkhead et al., 1994).

Although the prevailing patterns seem to support female choice for indirect benefits, a meta-analysis challenged this notion (Akçay and Roughgarden, 2007). Like within-pair paternity, extra-pair paternity requires two partners. Although still debated, there are clear benefits to females from extra-pair copulations (Evans and Magurran, 2000) and choosiness associated with that. But how about males? For males to engage in extra-pair matings is in agreement with the traditional view of sex roles where males are limited by the number of their matings and should consequently seek mating opportunities whenever possible. But are they also choosy when engaging in such matings and should they be (Lyu et al., 2017)? The only theoretical model relevant to the topic finds that male mate choice for costly female signals can indeed evolve and can, interestingly, be cyclical (Lyu et al., 2017). The signals evolved in females could be egg color, which has been suggested as correlated with female quality (Moreno and Osorno, 2003). This would, of course be limited to species with visible eggs.

In humans extra-pair paternity is also well documented but seems to be less common, with an average of 2% of offspring resulting from extra-pair copulations (Grebe and Drea, 2018). Note that extra-pair copulations are a mechanism for both genders to thwart potentially negative effects of arranged relationships (Grebe and Drea, 2018) (Chapter 2). But clearly, the costs and benefits of extra-pair matings are different for women and men. Also, the genders differ in the choosiness and the criteria used (Buss, 2015). Men interested in short-term matings (at least potentially extra-pair matings) assessed potential partners less based on facial characteristics, but more based on their body, supposedly to assess fertility (Confer et al., 2010). Overall, mate choice relative to extra-pair matings is not well understood, and especially male mate choice is barely documented.

4.7 Short summary

Females clearly differ in many characteristics that could be relevant to males evaluating females—just like males. As in female choice, there is solid evidence that males select mating partners for direct benefits, but the majority of studies focus on fecundity. Fecundity is clearly important, but male mating preferences for other traits must be important too. Much more work is needed in these areas. Male mate choice for indirect benefits remains largely unexplored, as is male mate choice of compatible females. Adaptive adjustment of male ejaculates emerges as an interesting area. Most importantly, the question remains if male mate choice can drive the evolution of female ornaments?

4.8 Further reading

The comprehensive book on mate choice by Rosenthal is a great text to follow up on the general topic covered in this chapter (Rosenthal, 2017).

4.9 References

AKÇAY, E. & ROUGHGARDEN, J. 2007. Extra-pair paternity in birds: review of the genetic benefits. *Evolutionary Ecology Research*, 9, 855.

AMLACHER, J. & DUGATKIN, L. A. 2005. Preference for older over younger models during mate-choice copying in young guppies. *Ethology Ecology & Evolution*, 17, 161–9.

ANDERSSON, J., BORG-KARLSON, A.-K., & WIKLUND, C. 2000. Sexual cooperation and conflict in butterflies: a male-transferred anti-aphrodisiac reduces harassment of recently mated females. *Proceedings of the Royal Society of London—Series B: Biological Sciences*, 267, 1271–5.

ARRIAGA, L. R. & SCHLUPP, I. 2013. Poeciliid male mate preference is influenced by female size but not by fecundity. *Peerj*, 1, e140.

ASPBURY, A.S. & GABOR, C.R. 2004a. Discriminating males alter sperm production between species. *Proceedings of the National Academy of Sciences of the United States of America*, 101, 15970–73.

ASPBURY, A.S. & GABOR, C.R. 2004b. Differential sperm priming by male sailfin mollies (*Poecilia latipinna*): effects of female and male size. *Ethology*, 110, 193–202.

BAKKER, T. C. M., KÜNZLER, R., & MAZZI, D. 1999. Condition-related mate choice in sticklebacks. *Nature*, 401, 234.

BARBER, I., HOARE, D., & KRAUSE, J. 2000. Effects of parasites on fish behaviour: a review and evolutionary perspective. *Reviews in Fish Biology and Fisheries*, 10, 131–65.

BARRY, K. L. & KOKKO, H. 2010. Male mate choice: why sequential choice can make its evolution difficult. *Animal Behaviour*, 80, 163–9.

BARUFFALDI, L. & COSTA, F. G. 2014. Male reproductive decision is constrained by sex pheromones produced by females. *Behaviour*, 151, 465–77.

BATEMAN, P. W. & FERGUSON, J. 2004. Male mate choice in the Botswana armoured ground cricket *Acanthoplus discoidalis* (Orthoptera: Tettigoniidae; Hetrodinae). Can, and how, do males judge female mating history? *Journal of Zoology*, 262, 305–9.

BELTRAN-BECH, S. & RICHARD, F.-J. 2014. Impact of infection on mate choice. *Animal Behaviour*, 90, 159–70.

BERSON, J. D. & SIMMONS, L. W. 2019. Female cuticular hydrocarbons can signal indirect fecundity benefits in an insect. *Evolution*, 73, 982–9.

BILBO, S. D., KLEIN, S. L., DEVRIES, A. C., & NELSON, R. J. 1999. Lipopolysaccharide facilitates partner preference behaviors in female prairie voles. *Physiology & Behavior*, 68, 151–6.

BIRKHEAD, T. & HUNTER, F. 1990. Mechanisms of sperm competition. *Trends in Ecology & Evolution*, 5, 48–52.

BIRKHEAD, T. R., HOSKEN, D. J., & PITNICK, S. S. 2008. *Sperm Biology: An Evolutionary Perspective*. Amsterdam: Elsevier.

BIRKHEAD, T. R. & MONTGOMERIE, R. 2008. Three centuries of sperm research. *In*: BIRKHEAD, T. R., HOSKEN, D. J., & PITNICK, S. S. (eds.) *Sperm Biology: An Evolutionary Perspective*. Amsterdam: Elsevier.

BIRKHEAD, T. R., VEIGA, J. P., & MØLLER, A. 1994. Male sperm reserves and copulation behaviour in the house sparrow, *Passer domesticus*. *Proceedings of the Royal Society of London—Series B: Biological Sciences*, 256, 247–51.

BURAND, J. P. & TAN, W. 2006. Mate preference and mating behavior of male *Helicoverpa zea* (Lepidoptera: Noctuidae) infected with the sexually transmitted insect virus Hz-2V. *Annals of the Entomological Society of America*, 99, 969–73.

BUSS, D. M. 2007. The evolution of human mating. *Acta Psychologica Sinica*, 39, 502–12.

BUSS, D. 2015. *Evolutionary Psychology*. Abingdon: Routledge.

BUSS, D. M. & SCHMITT, D. P. 2019. Mate preferences and their behavioral manifestations. *Annual Review of Psychology*, 70, 77–110.

BYERS, J., HEBETS, E., & PODOS, J. 2010. Female mate choice based upon male motor performance. *Animal Behaviour*, 79, 771–8.

CARAZO, P., SANCHEZ, E., FONT, E., & DESFILIS, E. 2004. Chemosensory cues allow male *Tenebrio molitor* beetles to assess the reproductive status of potential mates. *Animal Behaviour*, 68, 123–9.

CARDOZO, G., DEVIGILI, A., ANTONELLI, P., & PILASTRO, A. 2020. Female sperm storage mediates post-copulatory costs and benefits of ejaculate anticipatory plasticity in the guppy. *Journal of Evolutionary Biology*, 33, 1294–305.

CLERMONT, J., RÉALE, D., & GIROUX, J.-F. 2019. Similarity in nest defense intensity in Canada goose pairs. *Behavioral Ecology and Sociobiology*, 73, 108.

CONFER, J. C., PERILLOUX, C., & BUSS, D. M. 2010. More than just a pretty face: men's priority shifts toward bodily attractiveness in short-term versus long-term mating contexts. *Evolution and Human Behavior*, 31, 348–53.

CONROY-BEAM, D. & BUSS, D. M. 2019. Why is age so important in human mating? Evolved age preferences and their influences on multiple mating behaviors. *Evolutionary Behavioral Sciences*, 13, 127.

DARWIN, C. 1871. *The Descent of Man*. London: John Murray.

DEATON, R. 2009. Effects of a parasitic nematode on male mate choice in a livebearing fish with a coercive mating system (western mosquitofish, *Gambusia affinis*). *Behavioural Processes*, 80, 1–6.

DEWSBURY, D. A. 1982. Ejaculate cost and male choice. *The American Naturalist*, 119, 601–10.

EBERHARD, W. G. 1996. *Female Control: Sexual Selection by Cryptic Female Choice*. Princeton, NJ: Princeton University Press.

ENGQVIST, L. & REINHOLD, K. 2006. Theoretical influence of female mating status and remating propensity on male sperm allocation patterns. *Journal of Evolutionary Biology*, 19, 1448–58.

EVANS, J. P. & MAGURRAN, A. E. 2000. Multiple benefits of multiple mating in guppies. *Proceedings of the National Academy of Sciences of the United States of America*, 97, 10074–6.

FERNANDEZ, A. M., MUÑOZ-REYES, J. A., & DUFEY, M. 2014. BMI, age, mate value, and intrasexual competition in Chilean women. *Current Psychology*, 33, 435–50.

FIRMAN, R. C., GASPARINI, C., MANIER, M. K., & PIZZARI, T. 2017. Postmating female control: 20 years of cryptic female choice. *Trends in Ecology & Evolution*, 32, 368–82.

FITZPATRICK, J. L., WILLIS, C., DEVIGILI, A., YOUNG, A., CARROLL, M., HUNTER, H. R., & BRISON, D. R. 2020. Chemical signals from eggs facilitate cryptic female choice in humans. *Proceedings of the Royal Society of London—Series B: Biological Sciences*, 287, 20200805.

GANGESTAD, S. W. & BUSS, D. M. 1993. Pathogen prevalence and human mate preferences. *Ethology and Sociobiology*, 14, 89–96.

GASKETT, A. 2007. Spider sex pheromones: emission, reception, structures, and functions. *Biological Reviews*, 82, 27–48.

GIRNDT, A., CHNG, C. W. T., BURKE, T., & SCHROEDER, J. 2018. Male age is associated with extra-pair paternity, but not with extra-pair mating behaviour. *Scientific Reports*, 8, 8378.

GIRNDT, A., COCKBURN, G., SÁNCHEZ-TÓJAR, A., HERTEL, M., BURKE, T., & SCHROEDER, J. 2019. Male age and its association with reproductive traits in captive and wild house sparrows. *Journal of Evolutionary Biology*, 32, 1432–43.

GOETZ, A. T., SHACKELFORD, T. K., PLATEK, S. M., STARRATT, V. G., & MCKIBBIN, W. F. 2007. Sperm competition in humans: implications for male sexual psychology, physiology, anatomy, and behavior. *Annual Review of Sex Research*, 18, 1–22.

GREBE, N. M. & DREA, C. M. 2018. *Human Sexuality. In:* SHACKELFORD, T.K. & WEEKES-SHACKELFORD, V. A. (eds.) *Encyclopedia of Evolutionary Psychological Science*. Gland, Switzerland: Springer International, pp. 1–14.

GRIFFITH, S. C., OWENS, I. P. F., & THUMAN, K. A. 2008. Extra pair paternity in birds: a review of interspecific variation and adaptive function. *Molecular Ecology*, 11, 2195–212.

GUEVARA-FIORE, P. & ENDLER, J. A. 2018. Female receptivity affects subsequent mating effort and mate choice in male guppies. *Animal Behaviour*, 140, 73–9.

GUEVARA-FIORE, P., SKINNER, A., & WATT, P. J. 2009. Do male guppies distinguish virgin females from recently mated ones? *Animal Behaviour*, 77, 425–31.

HALL, M. L. 2004. A review of hypotheses for the functions of avian duetting. *Behavioral Ecology and Sociobiology*, 55, 415–30.

HALL, M. L. 2009. A review of vocal duetting in birds. *Advances in the Study of Behavior*, 40, 67–121.

HAMILTON, W. D. & ZUK, M. 1982. Heritable true fitness and bright birds: a role for parasites. *Science*, 218, 384–7.

HEBETS, E. A., WESSON, J., & SHAMBLE, P. S. 2008. Diet influences mate choice selectivity in adult female wolf spiders. *Animal Behaviour*, 76, 355–63.

HEINS, D. & BAKER, J. 2008. The stickleback–*Schistocephalus* host–parasite system as a model for understanding the effect of a macroparasite on host reproduction. *Behaviour*, 145, 625–45.

HEINS, D. C., ULINSKI, B., JOHNSON, J., & BAKER, J. A. 2004. Effect of the cestode macroparasite *Schistocephalus pungitii* on the reproductive success of ninespine stickleback, *Pungitius pungitius*. *Canadian Journal of Zoology*, 82, 1731–7.

HUGHES, S. M., HARRISON, M. A., & GALLUP JR, G. G. 2007. Sex differences in romantic kissing among college students: an evolutionary perspective. *Evolutionary Psychology*, 5, 147470490700500310.

HUNT, J., BROOKS, R., & JENNIONS, M. D. 2005. Female mate choice as a condition-dependent life-history trait. *The American Naturalist*, 166, 79–92.

JEANNERAT, E., MARTI, E., BERNEY, C., JANETT, F., BOLLWEIN, H., SIEME, H., BURGER, D., & WEDEKIND, C. 2018. Stallion semen quality depends on major histocompatibility complex matching to teaser mare. *Molecular Ecology*, 27, 1025–35.

JOHNSON, S. L., DUNLEAVY, J., GEMMELL, N. J., & NAKAGAWA, S. 2015. Consistent age-dependent declines in human semen quality: a systematic review and meta-analysis. *Ageing Research Reviews*, 19, 22–33.

JONES, T. M. & ELGAR, M. A. 2004. The role of male age, sperm age and mating history on fecundity and fertilization success in the hide beetle. *Proceedings of the Royal Society of London—Series B: Biological Sciences*, 271, 1311–18.

KEMPENAERS, B., VERHEYEN, G. R., & DHONDI, A. A. 1997. Extrapair paternity in the blue tit (*Parus caeruleus*): female choice, male charateristics, and offspring quality. *Behavioral Ecology*, 8, 481–92.

KEMPENAERS, B., VERHEYEN, G. R., VAN DEN BROECK, M., BURKE, T., VAN BROECKHOVEN, C., & DHONDT, A. 1992. Extra-pair paternity results from female preference for high-quality males in the blue tit. *Nature*, 357, 494.

KENNEDY, C., Endler, J., Poynton, S., & McMinn, H. 1987. Parasite load predicts mate choice in guppies. *Behavioral Ecology & Sociobiology*, 21, 291–6.

KING, B., SAPORITO, K., ELLISON, J., & BRATZKE, R. 2005. Unattractiveness of mated females to males in the parasitoid wasp *Spalangia endius*. *Behavioral Ecology and Sociobiology*, 57, 350–6.

KNELL, R. J. & WEBBERLEY, K. M. 2004. Sexually transmitted diseases of insects: distribution, evolution, ecology and host behaviour. *Biological Reviews*, 79, 557–81.

KOTUSZ, J., POPIOŁEK, M., DROZD, P., DE GELAS, K., ŠLECHTOVÁ, V., & JANKO, K. 2014. Role of parasite load and differential habitat preferences in maintaining the coexistence of sexual and asexual competitors in fish of the *Cobitis taenia* hybrid complex. *Biological Journal of the Linnean Society*, 113, 220–35.

LARSDOTTER MELLSTRÖM, H., & WIKLUND, C. 2009. Males use sex pheromone assessment to tailor ejaculates to risk of sperm competition in a butterfly. *Behavioral Ecology*, 20, 1147–51.

LARUE, K. M., CLEMENS, J., BERMAN, G. J., & MURTHY, M. 2015. Acoustic duetting in *Drosophila virilis* relies on the integration of auditory and tactile signals. *Elife*, 4, e07277.

LIVELY, C. M. 2010. A review of Red Queen models for the persistence of obligate sexual reproduction. *Journal of Heredity*, 101, S13–20.

LÓPEZ, S. 1999. Parasitized female guppies do not prefer showy males. *Animal Behaviour*, 57, 1129–34.

LYU, N., SERVEDIO, M. R., LLOYD, H., & SUN, Y. H. 2017. The evolution of postpairing male mate choice. *Evolution*, 71, 1465–77.

MANSER, A., KÖNIG, B., & LINDHOLM, A. 2015. Female house mice avoid fertilization by t haplotype incompatible males in a mate choice experiment. *Journal of Evolutionary Biology*, 28, 54–64.

MANSER, A., LINDHOLM, A. K., & WEISSING, F. J. 2017. The evolution of costly mate choice against segregation distorters. *Evolution*, 71, 2817–28.

MAZZI, D. 2004. Parasites make male pipefish careless. *Journal of Evolutionary Biology*, 17, 519–27.

MENDOZA-CUENCA, L. & MACIAS-ORDONEZ, R. 2010. Female asynchrony may drive disruptive sexual selection on male mating phenotypes in a *Heliconius* butterfly. *Behavioral Ecology*, 21, 144–52.

MESTON, C. M. & BUSS, D. M. 2009. *Why Women Have Sex: Understanding Sexual Motivations from Adventure to Revenge (and Everything in between)*. London: Macmillan.

MILINSKI, M. 2006. The major histocompatibility complex, sexual selection, and mate choice. *Annual Review of Ecology Evolution and Systematics*, 37, 159–86.

MOORE, J. 2002. *Parasites and the Behavior of Animals*. Oxford: Oxford University.

MORENO, J. & OSORNO, J. L. 2003. Avian egg colour and sexual selection: does eggshell pigmentation reflect female condition and genetic quality? *Ecology Letters*, 6, 803–6.

NEIMAN, M. & KOSKELLA, B. 2010. Sex and the Red Queen. *In*: SCHÖN, I., MARTENS, K., & VAN DIJK, P. (eds.) *Lost Sex: The Evolutionary Biology of Parthenogenesis*. Heidelberg: Springer.

PARKER, G. 1984. Sperm competition and the evolution of animal mating strategies. *In*: SMITH, R. L. (ed.) *Sperm Competition and the Evolution of Animal Mating Systems*. Amsterdam: Elsevier, pp. 1–60.

PARKER, G. A. 1970. Sperm competition and its evolutionary consequences in the insects. *Biological Reviews*, 45, 525–67.

PFENNIG, K. S. 2007. Facultative mate choice drives adaptive hybridization. *Science*, 318, 965–7.

PLATH, M., HEUBEL, K. U., GARCIA DE LEON, F. J., & SCHLUPP, I. 2005. Cave molly females (*Poecilia mexicana*, Poeciliidae, Teleostei) like well-fed males. *Behavioural Ecology and Sociobiology*, 58, 144–51.

PLATH, M., RICHTER, S., TIEDEMANN, R., & SCHLUPP, I. 2008. Male fish deceive competitors about mating preferences. *Current Biology*, 18, 1138–41.

PLATH, M. & SCHLUPP, I. 2008. Misleading mollies: the effect of an audience on the expression of mating preferences. *Communicative & Integrative Biology*, 1, 199–203.

RĘK, P. & MAGRATH, R. D. 2016. Multimodal duetting in magpie-larks: how do vocal and visual components contribute to a cooperative signal's function? *Animal Behaviour,* 117, 35–42.

RHAINDS, M. 2019. Ecology of female mating failure/ lifelong virginity: a review of causal mechanisms in insects and arachnids. *Entomologia Experimentalis et Applicata,* 167, 73–84.

RIECHERT, S. E. & SINGER, F. D. 1995. Investigation of potential male mate choice in a monogamous spider. *Animal Behaviour,* 49, 715–23.

RIESCH, R., PLATH, M., & SCHLUPP, I. 2010. Toxic hydrogen sulfide and dark caves: life-history adaptations in a livebearing fish (*Poecilia mexicana,* Poeciliidae). *Ecology,* 91, 1494–505.

RIESCH, R., REZNICK, D. N., PLATH, M., & SCHLUPP, I. 2016. Sex-specific local life-history adaptation in surface- and cave-dwelling Atlantic mollies (*Poecilia mexicana*). *Scientific Reports,* 6, 1–13.

RIESCH, R., SCHLUPP, I., & PLATH, M. 2008. Female sperm limitation in natural populations of a sexual/ asexual mating complex (*Poecilia latipinna, Poecilia formosa*). *Biology Letters,* 4, 266–9.

RIESCH, R., TOBLER, M., PLATH, M., & SCHLUPP, I. 2009. Offspring number in a livebearing fish (*Poecilia mexicana,* Poeciliidae): reduced fecundity and reduced plasticity in a population of cave mollies. *Environmental Biology of Fishes,* 84, 89–94.

ROBINSON, D.M., ASPBURY, A.S., & GABOR, C.R. 2008. Differential sperm expenditure by male sailfin mollies, *Poecilia latipinna,* in a unisexual–bisexual species complex and the influence of spermiation during mating. *Behavioral Ecology and Sociobiology,* 62, 705–11.

ROSENTHAL, G. G. 2017. *Mate Choice.* Princeton, NJ: Princeton University Press.

RYAN, M. J. & KEDDY-HECTOR, A. 1992. Directional patterns of female mate choice and the role of sensory biases. *American Naturalist,* 139, S4–35.

RYAN, M. J. & WAGNER, W. E. 1987. Asymmetries in mating preferences between species: female swordtails prefer heterospecific males. *Science,* 236, 595–7.

SCHLUPP, I. 2018. Male mate choice in livebearing fishes: an overview. *Current Zoology,* 64, 393–403.

SCHLUPP, I. & PLATH, M. 2005. Male mate choice and sperm allocation in a sexual/asexual mating complex of *Poecilia* (Poeciliidae, Teleostei). *Biology Letters,* 1, 169–71.

SCHLUPP, I. & RIESCH, R. 2011. Evolution of unisexual reproduction. *In:* SCHLUPP, I., PILASTRO, A., & EVANS, J. (eds.) *Ecology and Evolution of Poeciliid Fishes.* Chicago, IL: University of Chicago Press, pp. 50–8.

SEIBT, U. & WICKLER, W. 1977. Duettieren als Revier-Anzeige bei Vögeln. *Zeitschrift für Tierpsychologie,* 43, 180–7.

SHINE, R. 1988. The evolution of large body size in females: a critique of Darwin's "fecundity advantage" model. *The American Naturalist,* 131, 124–31.

SINGH, D. 1993. Adaptive significance of female physical attractiveness: role of waist-to-hip ratio. *Journal of Personality and Social Psychology,* 65, 293.

SIROT, L. K. 2019. Modulation of seminal fluid molecules by males and females. *Current Opinion in Insect Science,* 35, 109–16.

SIROT, L. K., FINDLAY, G. D., SITNIK, J. L., FRASHERI, D., AVILA, F. W., & WOLFNER, M. F. 2014. Molecular characterization and evolution of a gene family encoding both female- and male-specific reproductive proteins in *Drosophila*. *Molecular Biology and Evolution,* 31, 1554–67.

SWINEY, K. M., LONG, W. C., ECKERT, G. L., & KRUSE, G. H. 2012. Red king crab, *Paralithodes camtschaticus,* size-fecundity relationship, and interannual and seasonal variability in fecundity. *Journal of Shellfish Research,* 31, 925–33.

THOMAS, F., RENAUD, F., & CEZILLY, F. 1996. Assortative pairing by parasitic prevalence in *Gammarus insensibilis* (Amphipoda): patterns and processes. *Animal Behaviour,* 52, 683–90.

THOMAS, M.L. & SIMMONS, L.W. 2007. Male crickets adjust the viability of their sperm in response to female mating status. *The American Naturalist,* 170, 190–5.

THOMAS, M. L. & SIMMONS, L. W. 2010. Cuticular hydrocarbons influence female attractiveness to males in the Australian field cricket, *Teleogryllus oceanicus.* *Journal of Evolutionary Biology,* 23, 707–14.

TOBLER, M. & SCHLUPP, I. 2008. Expanding the horizon: the Red Queen and potential alternatives. *Canadian Journal of Zoology,* 86, 765–73.

TOVÉE, M. J., MAISEY, D. S., EMERY, J. L., & CORNELISSEN, P. L. 1999. Visual cues to female physical attractiveness. *Proceedings of the Royal Society of London—Series B: Biological Sciences,* 266, 211–18.

UHIA, E. & RIVERA, A. C. 2005. Male damselflies detect female mating status: importance for postcopulatory sexual selection. *Animal Behaviour,* 69, 797–804.

VAN VALEN, L. 1973. A new evolutionary law. *Evolutionary Theory,* 1, 1–30.

VEENENDAAL, M. V., PAINTER, R. C., DE ROOIJ, S. R., BOSSUYT, P. M., VAN DER POST, J. A., GLUCKMAN, P. D., HANSON, M. A., & ROSEBOOM, T. J. 2013. Transgenerational effects of prenatal exposure to the 1944–45 Dutch famine. *BJOG: An International Journal of Obstetrics & Gynaecology,* 120, 548–54.

WAGNER, W. E. J. 1998. Measuring female mating preferences. *Animal Behaviour,* 55, 1029–42.

WEDEKIND, C. 2018. A predicted interaction between odour pleasantness and intensity provides evidence for major histocompatibility complex social signalling in women. *Proceedings of the Royal Society of London—Series B: Biological Sciences,* 285, 20172714.

WEDEKIND, C. & FÜRI, S. 1997. Body odour preferences in men and women: do they aim for specific MHC combinations or simply heterozygosity? *Proceedings of the Royal Society of London—Series B: Biological Sciences*, 264, 1471–9.

WEDELL, N. & COOK, P. A. 1999. Butterflies tailor their ejaculate in response to sperm competition risk and intensity. *Proceedings of the Royal Society of London—Series B: Biological Sciences*, 266, 1033–9.

WEDELL, N., GAGE, M. J., & PARKER, G. A. 2002. Sperm competition, male prudence and sperm-limited females. *Trends in Ecology & Evolution*, 17, 313–20.

WEDELL, N. & RITCHIE, M. G. 2004. Male age, mating status and nuptial gift quality in a bushcricket. *Animal Behaviour*, 67, 1059–65.

WEGNER, K. M., KALBE, M., KURTZ, J., REUSCH, T. B. H., & MILINSKI, M. 2003. Parasite selection for immunogenetic optimality. *Science (Washington D C)*, 301, 1343.

YIĞIT, A., SEVGILI, H., & ÖZDEMIR, H. 2019. Male bushcrickets in female-biased environment allocate fewer sperm per ejaculation. *Entomological News*, 128, 393–403.

ZUK, M. 1992. The role of parasites in sexual selection—current evidence and future directions. *Advances in the Study of Behavior*, 21, 39–68.

CHAPTER 5

Male Investment and Male Choice

5.1 Brief outline of the chapter

In this chapter I revisit the role that male investment might play in the evolution of male mate choice. Simply put, with more male investment in reproduction one might expect them to be more selective regarding with whom they mate. If the male investment is larger than the female investment sex roles are expected to flip. This is, however, not common. But even if males almost never invest as much or more than females, in many species they do invest somewhat, or even heavily. Should this lead to choosiness in males? I will provide a few examples where choosiness may be linked to investment in males.

5.2 Introduction

Male investment is at the core of understanding male mate choice. This is because we view the evolution of female choice as relative to female investment in their gametes, and consequently an important pathway toward male mate choice is via investment in the offspring. Of course, other pathways for the evolution of male mate choice are known to be relevant, too, such as variation in female quality (Chapter 4) and a female-biased sex ratio (Chapter 6). Nonetheless, it seems that male investment is important because of the importance of sex roles. Indeed, as discussed in Chapter 1 and 2, the whole concept of sex roles rests mainly on the huge differential in investment in gametes. And consequently, a key argument put forward against more relevance of male mate choice is based on the traditional view of sex roles (Bateman, 1948; Trivers, 1972). In essence this is an argument based on the economy and ecology of investment and

posits that whoever invests the most is choosing. Because of the strong early investment by females, which is often continued, it is both unlikely and difficult for males to catch up with and compensate for female investment and even more unlikely for them to surpass it. One of the issues with sex roles, however, is that they are viewed as mostly binary. In my opinion this introduces a dichotomy that may not actually be that important. Essentially a binary view of sex roles makes it seem as if either females or males choose. However, female and male choice can occur simultaneously, as discussed in the rich literature on mutual mate choice (Johnstone et al., 1996; Johnstone, 1997; Kokko and Johnstone, 2002) (Chapter 9). I am not arguing that the differential in investment should not lead to female choice, however, if males do invest somewhat, should they also evolve choosiness, especially if they can invest differentially into some mating partners or offspring? Should choosiness be proportional to the investment in some way? If a male—for example—invests 60% of what a female invests, should his choosiness—maybe measured as strength of preferences—be 60% of female choosiness? Furthermore, what do we reasonably count as investment? Based on the assumption that investment in reproduction may be determining choosiness we need to tally such investments by males and females. More importantly, I think that even limited male investment predicts some choosiness in males. It should be noted that this is different from the concept that males choose based on differences in female quality (Chapter 4) or owing to sex ratio (operational sex ratio [OSR] or adult sex ratio [ASR]) (Chapter 6), but these three mechanisms are not mutually exclusive (Chapters 6 and 9). Parental investment and

Male Choice, Female Competition, and Female Ornaments in Sexual Selection. Ingo Schlupp, Oxford University Press (2021). © Ingo Schlupp (2021).
DOI: 10.1093/oso/9780198818946.003.0005

OSR/ASR are connected, of course (Kokko and Jennions, 2008) as both concepts are related to sex roles. The pathways for the evolution of male mate choice based on these two concepts are disassociated from male investment in reproduction and Bateman gradients and should be able to evolve even in species with no male investment beyond sperm (Schlupp, 2018). In this chapter I briefly look at male investment not only in the offspring, but also in the mating partner and how they may be related to mate choice. And finally, I look at investments that are needed for a male to obtain matings and ask if they could predict choosiness.

Clearly, at first sight the notion that males invest very little in their gametes compared with females is obvious and correct (Bateman, 1948; Trivers, 1972). Yet, to be fair, this view has also been criticized (Gowaty et al., 2012) (Chapter 2). Obviously, however, comparing one egg with one sperm is hardly an appropriate comparison. If we agree that this is probably not the best metric, what should be included in the comparison? Quite clearly, males in many species invest considerably beyond the sperm cell. On the other hand, of course females also invest beyond the egg cell and in many species the imbalance persists, and females make a much bigger investment compared with males. Consequently, they evolve to be choosy. But I think a slightly more fine-grained view of forms of male investment would help us understand the potential for male investment in male mate choice.

For example, in humans, with the—sometimes limited—monogamy our species shows, women invest in a long pregnancy, followed by lactation and then a long time raising a child to adulthood. Men, on the other hand, start investing beyond the sperm once the pregnancy begins by investing in their partner and continue to invest in their children until they become independent. Men can also invest in a relationship before a reproductive event occurs. However, in men there are large differences in this investment.

Morally and legally, women and men are both expected to invest in their children (Chapter 2). This is expected by society and also codified in the law of many countries. As always, however, in humans, much of the documented variability between societies is under a strong cultural influence. This makes, however, an important point: behaviors are very individual and sometimes population parameters like median or mean values do not capture the cost–benefit matrix an individual may face. In humans we know about individual circumstances and how they influence behavior, but our attempts to understand behavior in other animals on a more individual basis are a little bit behind.

Generally speaking, it would be useful to be able to quantify such investments easily, but unfortunately the currencies of investment can be very different and may change over time. In mammals, early on the currency may be energy, but there is also time invested, opportunities missed, and—in humans—money and social capital invested. Large differences in the amount of investment along any of these axes exist in men, but very clearly most men make substantial investments in their relationship, partners, and offspring. How should that affect their behavior, especially their mate choice? We know from many studies in humans that women prefer men who show indicators that they will be good providers.

5.2.1 Investment prior to mating: does it count?

One aspect to bear in mind in this context is that males (or courters) can make investments before they engage in matings, or after they have mated with a partner. These two phases can be intertwined, but I find it helpful to consider them separately here. The flipside is that females (or choosers) may have to make decisions based on potential, not actual performance. Consequently, in humans, women choose based on indicators of male potential, and in birds nuptial feeding has been correlated with subsequent paternal behavior and females have been shown to base their preference on this (Wiggins and Morris, 1986), to mention but two examples. As a corollary, in humans understanding the emotion of jealousy has been framed in this context. Simply put, men are jealous when their partners have sex with other men because it increases the uncertainty of paternity. Women on the other hand are thought to be jealous when other women jeopardize their access to paternal investment (Oberzaucher and Grammer, 2009).

Since males of different species differ widely in both how and how much they invest in mating and offspring, there is a continuum in how much males invest from practically nothing in species where males only provide sperm (or an ejaculate in the case of internal fertilization) to males that invest a great deal, to the point that they become the primary source of care for the young. In some cases, this has led to what we call sex-role reversal (Chapter 2). However, in this section, I want to focus on the potential role of male investment in the behaviors and structures needed to obtain matings.

As we know now, most males produce many, yet relatively inexpensive gametes. When they deliver them, they typically bundle them into packages containing large numbers, often millions, of individual sperm cells. These sperm cells are accompanied by seminal fluid, and together they comprise the ejaculate. Clearly, an ejaculate is much more costly to produce than a single sperm cell, but in almost all cases the physiological cost for this is not nearly on the same level as an individual egg cell. There are interesting exceptions, though. Individual sperm is usually very small (Birkhead et al., 2008), but in some species of *Drosophila*, individual sperm have evolved to be very large. Consequently, such sperm is considered to be an ornament, and—probably based on this investment—males are choosy (Lüpold et al., 2010, 2016). Furthermore, in many insects, the packages or spermatophores to deliver the sperm can be very large and contain substances that are considered as direct benefits to either the female and/or the offspring (Sakaluk et al., 2019). These and other examples highlight that sperm is not always cheap and that males can be sperm limited. Especially in such cases high-quality males can evolve to be choosy, whereas low-quality males may drop out of the reproductive race completely. This is the case in some Australian bush crickets, where under harsh conditions only a few males are able to produce costly spermatophores (Chapter 6) (Gwynne and Simmons, 1990). This leads to a female-biased sex ratio, which in turn leads to male choice and female competition. But the underlying mechanism is male investment in spermatophores.

Sperm competition may be especially important in those situations where females seek matings with a few highly attractive males owing to their preferences converging on the same males, such as for example in leks (Bradbury et al., 1985; Höglund and Alatalo, 2014). Those males are likely to be sperm limited rather quickly, which in turn should result in those males being selective in whom they mate with.

As I said before, males invest in obtaining mating opportunities before they even come close to an actual mating. Many such traits—which we usually consider ornaments—have a fixed cost. Once the structure for an ornament is built, the investment has been completed and may have to be maintained, but often it does not need to be built all over again. Of course, signals can be seasonal, like the breeding coloration in birds, or nuptial coloration in fishes (Bradbury and Vehrencamp, 2011). Male deer, for example, invest heavily in antlers season after season and these antlers are important in female mate choice (Clutton-Brock et al., 1989), just as they are armaments (Berglund et al., 1996). One especially dynamic system with seemingly heavy investment in male ornaments is found in bower birds (Diamond, 1986), although Borgia (1993) argued that bowers are actually not very costly to males. Often viewed as an extended phenotype (Dawkins, 1982), the bowers built by males are nonetheless an impressive investment in ornaments that does not seem to lead to any choosiness in males, and only the females are selective (Borgia, 1995).

Overall, this highlights that the investment of males in matings is typically substantially higher than the investment in germ cells. However, these male investments are somewhat difficult to evaluate and measure. One could argue that investments in the infrastructure needed to obtain matings, like ornaments, should count because without that investment males would not obtain matings. Indeed, there are many species with polymorphisms in male ornaments, or, in other words, male strategies to obtain matings. To provide just one example, in sunfish some males are ornamented and court females, whereas other males are not ornamented and attempt to sneak copulations. These alternative strategies can be maintained in the population as balanced polymorphisms (Oliveira et al., 2008). How these alternative strategies affect male mate choice is not well known. Because males invest differently in ornaments that serve as signals to females, one would predict that males with a high investment might be choosier than males that make a small investment. They might, for example, be more likely to forgo matings with lower-quality females, especially if they are limited in some way, such as by the amount of sperm they can invest. This general idea has been tested at the species level in the fish genus *Limia*. These fishes are endemic to the Greater Antilles and species differ strongly in the degree of ornamentation (Figure 5.1). Spikes and colleagues (2021), compared species with uniform low investment in ornaments and species where some males show very high investment in ornaments. Ornamentation had already been independently classified by another study (Goldberg et al., 2019). The prediction that more ornamented males are choosier was not confirmed, but also males and females of this particular genus seem to be less choosy than almost all other livebearing fishes. Overall, however, there seems to be little evidence for male investment in ornaments being coupled with choosiness.

Figure 5.1 Differences in male size and investment in the sailfin molly (*Poecilia latipinna*). Similar patterns are also found in some species of *Limia* (photo credit: Ingo Schlupp).

Other investments by males might tell a different story: some relevant investments are not structural and can therefore be adjusted and provide a mechanism for males to exercise choice. One example of this is the composition of the ejaculate, which is usually interpreted in the context of sexual conflict (Torres-Vila and Jennions, 2005; Gwynne, 2008; South and Lewis, 2011; Lehmann, 2012, Lewis and South, 2012; Macartney et al., 2018; Sirot, 2019). There seems to be, however, substantial individual variation between individuals that might indicate some form of cryptic male mate choice (see Chapter 4). More work looking into the dynamic nature of cryptic female and male choice (Magris et al., 2017; Cardozo et al., 2020) would be useful. In particular, the female × male and male × male interactions in species with internal fertilization, such as guppies (*Poecilia reticulata*) and *Drosophila melanogaster*, are important emerging fields of research (Gasparini et al., 2012; Evans et al., 2013; Firman et al., 2017; Lüpold et al., 2020).

One problem with considering investments made prior to matings is that they are difficult or impossible to separate from investments in multiple other functions. For example, the muscle and ATP used to produce a motion pattern in the courtship of a bird, or mammal, can also be used in general locomotion and may be critically important for survival, not only courtship. On the other hand, the cost and effort invested in a nuptial gift or in feeding offspring is much more easily classified as such and measured. Another difficult currency to measure is time. Time used for courtship could also be used for foraging. This creates opportunity cost and significant trade-offs: a well-fed male that successfully avoids predators and parasites but never courts a female will nonetheless die as a well-fed, healthy virgin.

5.2.2 Male investment in offspring

Parental investment theory is based on a classical paper by Trivers (1972). Male investment usually comes as some form of paternal care. Males can feed offspring, protect or defend them in some way, manipulate the environment to their benefit, or contribute in any way to raising them. For example, in birds males can build a nest, feed their partners—either during courtship or during incubation—

incubate the eggs and offspring, defend them against predators, and provision the offspring.

Paternal care has evolved many times in many different taxa. It can be a shared responsibility with others, or the sole form of parental care. The time males spend caring for offspring can vary from a short fraction of their life to a longer commitment as found in some primates (Wright, 1990) including humans (*Homo sapiens*), where paternal care can span more than two decades. This is only one example for the many species where both females and males provide parental care, and in some cases of social animals even male helpers may provide care, as for example in Florida scrub jays, *Aphelocoma coerulescens* (Skutch, 1935; Woolfenden, 1975).

Male care is the only form of parental care in pipefishes (Vincent et al., 1992) and many other fishes, including cichlids (Blumer, 1979; Goodwin et al., 1998), but rarely in arthropods (Zeh and Smith, 1985; Tallamy, 2001). In pipefishes the reversed sex roles are correlated with male mate choice. But how about the many other cases where males also invest significantly in the offspring? Do males show signs of choosiness based on investment? One of the few better known examples from insects is found in heteroptera in the family Belostomatidae, including in the genus *Belostoma*, where males carry eggs on their back until they hatch (Ichikawa, 1989). It is known that females choose males (Smith, 2019), but also—in *Belostoma lutarium*—that males exercise preferences for heavier females (Thrasher et al., 2015). It seems that both sexes choose in this group.

In fishes the evolution of biparental, maternal, and paternal care has been investigated at different taxonomic levels. In cichlids, for example, paternal care in the form of mouthbrooding is hypothesized to have evolved only once, whereas maternal care has evolved many times (Goodwin et al., 1998). While the majority of fish species show no parental care at all, both maternal and paternal care have evolved several times in bony fishes (Benun Sutton and Wilson, 2019). Interestingly, evolutionary transitions from no care to either female or male care are relatively common, while transitions from male to female care or vice versa are very rare (Benun Sutton and Wilson, 2019). Another example for the latter would be fanning eggs in fishes. Other

behaviors in fishes include mouthbrooding, which is found in cichlids. In some frog species, males carry eggs and tadpoles. Motherhood is known to change the behavior of females and make them more risk averse. Interestingly, the opposite has been reported for fathers in a species of glass frog (*Hyalinobatrachium cappellei*), where males accept a higher mortality risk to attract additional females (Valencia-Aguilar et al., 2020). The question remains as to whether changes in parental care, as found several times in cichlids, are associated with changes in mate choice. In a study at the species level, this was found to be the case: in Galilee St Peter's fish, *Sarotherodon galilaeus*, a cichlid from Israel, both maternal and paternal mouthbrooding is found, and choosiness is associated with parental care and OSR (Balshine-Earn, 1996; Balshine-Earn and Earn, 1998).

As stated above, temporally, paternal investment in offspring can vary from lasting a short period of time during development to very long periods in long-lived species. In humans, for example, parental investment can happen over a very long part of the average life expectancy of an individual (Geary, 2015). According to the Centers for Disease Control and Prevention, in 2018 on average men in the United States had a life expectancy of 78.7 years. Assuming their children become independent at age 18, they may have spent a substantial part of their life investing in their children. Paternal investment occurs in many taxa, but its distribution is not even. For example, it is very rare in mammals, but relatively common in fishes. One suggested explanation for this pattern is that the partner that can leave after mating will often do so, thus saddling the partner who stays with the cost of additional investment in the offspring. In many fishes, for example, the male provides sperm after the eggs are laid by the female and she has left. In mammals, on the other hand, with internal fertilization and lactation representing massive female investment, it is easy for males to leave and not make additional investments in the offspring. This argument applies in principle to all livebearing organisms, not just mammals. For example, this argument makes it difficult to understand male investment in mammals, where males might benefit more from seeking additional mating opportunities, as compared with investing in offspring. In humans, ecological changes during the evolution of our species have been suggested to be important in the evolution of paternal care (Alger et al., 2020).

Furthermore, male investment is often fine-tuned and relative to the perceived probability of paternity. If paternity is highly likely for a male, investment should be high; if paternity is less likely, males may reduce their effort. In species with obligatory paternal investment, males that decrease the investment too much risk a total loss of their investment. Both females and males should adjust their investment in offspring based on the likelihood that they are the actual parents. For females this is virtually certain if they give birth to live young. In egg-laying species, however, even females cannot be certain of maternity if and when they care for their eggs. In many species of fishes and birds, for example, females lay eggs in the nests of other females, a behavior called egg-dumping. In almost all species fatherhood is always uncertain.

In humans, phenotype matching can be used to assess paternity. Children who look and smell like their putative father in a polygynous culture in Senegal receive more investment and are healthier (Alvergne et al., 2009). The likelihood of paternity can also influence the investment by grandfathers. Whereas maternal grandmothers can be certain that their grandchildren are actually closely related to them, the same is not true for paternal grandfathers. They are facing paternity uncertainty twice. Very few species show grandpaternal investment, but they would be worth looking at in terms of differences in investment. This uncertainty might prevent the evolution of male investment, but models focusing on ecological conditions support the evolution of male investment in humans (Alger et al., 2020).

If male investment in mating is a continuum, why does it often look like (or is presented as if) a binary situation, with one sex being the limited and the other the limiting sex? If males invest, they could benefit from selecting a female partner who is also investing. If a male invests strongly in the offspring, this makes him a desired mate, and at the same time this may turn a high-quality male into a rare commodity, which allows him to benefit from being choosy. In those cases where both males and

females can benefit from choosing, mutual mate choice may evolve. Of course, and to say it again, paternal investment is only one path to the evolution of male mate choice, with OSR and differences in female quality being the others. One question that in my opinion could benefit from more research is whether male choosiness is in any way scaling with male investment. This seems to be the case in humans, but more examples would be helpful.

5.2.3 Male investment in their partner

In addition to directly investing in their offspring, males can invest in the mother of their offspring. Such investment is much more indirect, and more prone to doubts regarding paternity, but this type of investment is very widespread and will often benefit the offspring sired by the male in question. This type of investment, however, allows females to receive an investment, often in the form of a nuptial gift, without awarding paternity to the males providing the gift. This can lead to intensive sexual conflict.

Males in many species feed their mating partners, either during courtship with a nuptial gift or later, during maternal care. The opposite, females feeding (or caring for) a mating partner, is rare, even in sex-role reversed species. Maybe the most extreme form is found in sexual cannibalism, where the male becomes the nuptial gift for the female (Andrade and Kasumovic, 2005). In other systems, males can present food items or a large spermatophore to the female (Sakaluk et al., 2019). Such nuptial gifts may contain nutritious substances, amino acids, water, or other components that help the mother and are sometimes passed on to the offspring (Figure 5.2). They are sometimes used to manipulate females, giving rise to sexual conflict. Interestingly females sometimes reject nuptial gifts as a mechanism to exercise female choice. Because the time she spends consuming the nuptial gift is correlated with the amount of sperm transferred, females may have to reject the gift to avoid inseminations with nonpreferred males (Sakaluk et al., 2019). Interestingly, nuptial gifts have evolved many times independently.

Nuptial gifts are widespread in insects but are also found in other taxa. In many species of birds,

Figure 5.2 Male decorated cricket, *Gryllodes sigillatus* (bottom), transferring a spermatophore to a female (top) during copulation, which includes a spermatophylax provided as a nuptial gift (photo credit: David Funk).

the male feeds the partner in addition to feeding the offspring.

5.3 Male investment is relative

Male investment can be modified based on the perceived probability of paternity. Similarly, females can adjust their effort when maternity is uncertain. This happens, for example, in the burying beetle, *Nicrophorus vespilloides* (Richardson and Smiseth, 2020). Females sometimes share the same carcass but cannot recognize their own offspring. Thus, when co-breeding they lay bigger and more eggs, but provide less care. In males, adjustment of paternal care relative to the probability of paternity has been shown in many birds. One of the many examples is the collared flycatcher, *Ficedula albicollis*, where males were experimentally manipulated (by removing them from the nest) to experience reduced paternity. This was followed by reduced male effort (Sheldon and Ellegren, 1998).

5.4 Life history tradeoffs and future reproduction

A hallmark of Trivers' parental investment theory and the parent offspring conflict theory (Trivers, 1972, 1974) is that investment in any given current offspring usually comes at a cost of future reproduction or present siblings, which can trigger several conflicts. Parents should therefore be pru-

dent about the optimal investment in any given offspring.

5.5 Three examples of male investment and male mate choice

I want to end this section with three examples of male investment that have led to the evolution of male mate choice, but to different degrees. In the first example, giant water bugs of the genus *Belostoma*, males appear to make significant investments, yet they show limited male mate choice. In the second example, sticklebacks, males also invest strongly, and they exercise clear choice, as do females. This example also highlights mutual mate choice. Finally, pipefishes show strong investment in parental care, and in this case, male choice has emerged as the almost exclusive mating behavior and we consider the mating system sex-role reversed. With this final example, I will also highlight the potential role of the OSR in the evolution of mate choice.

First to giant water bugs. Parental care has evolved several times in insects, but overall it is very rare (Smith, 1997). It is found, for example, in assassin bugs (Thomas and Manica, 2005), but also in giant water bugs of the genus *Belostoma*, where male mate choice has been detected in at least one species, but has not been reported in others. There are several potential reasons for this. Giant water bugs are relatively large aquatic heteropterans and prey on various other species, including snails (Crowl and Alexander Jr, 1989; Armúa de Reyes and Estévez, 2006) and fish (Crowl and Alexander Jr, 1989; Tobler et al., 2007; Plath et al., 2011,). Males carry eggs on their back, thus providing paternal care. In *Belostoma lutarium* a species from North America, this is associated with male mate choice, although it appears to be relatively weak. In other species of *Belostoma*, such as *Belostoma flumineum* also from North America, only female choice has been reported (Kight and Kruse, 1992; Kight et al., 2011). Although it is possible that male mate choice still awaits detection in more species of *Belostoma*, it is possible that the ecology and partly the OSR prevent male mate choice from occurring. This is probably because, despite male parental care, the sex ratio still seems to be male-biased because females need a relatively long time to produce clutches (Kruse, 1990) and there is intense sexual conflict (Härdling and Kaitala, 2001), as modeled for the golden egg bug (*Phyllomorpha laciniata*). I suggest that an experimental approach creating a more female-biased OSR might push this system to showing male choice (Chapter 6).

The second example I want to examine in this context is mutual mate choice in sticklebacks. This group of fishes has long been used as models in animal behavior (Tinbergen, 1952) and ecology and evolution (Wootton, 1984), and is known for both parental care and male choice. I introduced sticklebacks back in Chapter 3. In several species in this group males provide intensive paternal care, and in the three-spined stickleback (*Gasterosteus aculeatus*) males show a preference for larger females (Rowland, 1982; Rowland and Sevenster, 1985; Rowland, 1989; Kraak and Bakker, 1998; Candolin and Salesto, 2009). Furthermore, males prefer ornamented females (Rowland et al., 1991; Bakker and Rowland, 1995; Yong et al., 2018). Together, this is an example of mutual mate choice, with both sexes showing distinct preferences (Kraak and Bakker, 1998): males for larger and more fecund females and females for males that are redder. Males often breed only once and hence a more balanced OSR seems possible, or OSRs—which are theoretically predicted to be male-biased—vary over the season sufficiently to allow male mate choice to evolve (Smith et al., 1995).

Finally, in pipefishes, as in several other species that are considered sex-role reversed, male investment surpasses female investment and males become choosy, while females compete for males. For pipefishes, the Bateman gradient for males and females should be reversed and the gradient should be steeper for females. While this seems to be generally true, in a study on dusky pipefish (*Syngnathus floridae*), Mobley and Jones found that the gradients for males and females were remarkably similar (Mobley and Jones, 2013). This, together with some doubt cast on Bateman's results in general (Gowaty et al., 2012), presents a challenge for the role of investment in shaping mate choice. As in *Belostoma* and sticklebacks, other factors may be important too. One of them could be the OSR again. In one of the best studied pipefish species, *Syngnathus typhle* (Figure 5.3), from Scandinavia, the OSR changes over the breeding season, but remains essentially female-biased. Early in the season, however, it can be even, or male-biased. Over the season however, because males take so long to brood a clutch, the OSR becomes strongly female-biased (Vincent et al., 1994). In another pipefish mating system the sex ratio is also influential: in *Corythoichthys haematopterus*, males and females have relatively similar potential reproductive rates, but monogamy and a female-biased sex ratio lead to sex-role reversal and female competition for males (Sogabe and Yanagisawa, 2007). Finally, the black-striped pipefish, *Syngnathus abaster*, a widespread but southern species, was found to respond to changes in the sex ratio in a rather fine-tuned way and move back and forth from male-like to female-like behavior (Silva et al., 2010). Combined, these examples suggest that in addition to investment in the offspring, the sex ratio plays an important role. I find it especially intriguing that there seem to be cases where—either

Figure 5.3 Courting pair of broadnosed pipefish (*Syngnathus typhle*) (photo credit: Anders Berglund).

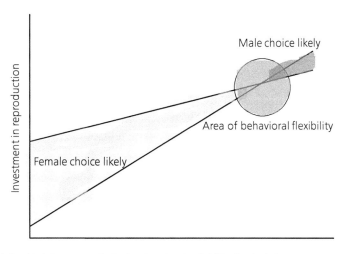

Figure 5.4 Hypothetical relationship between reproductive investment and probability of mate choice.

based on investment or sex ratio—species, populations, or individuals can alternate between male-like and female-like behavior. This is depicted as the area of behavioral flexibility in Figure 5.4. I will expand on the role of the sex ratio in Chapter 6. My suggestion is quite similar to the switch point theorem (Gowaty and Hubbell, 2009; Ah-King and Gowaty, 2016), which predicts flexibility in mating strategies based on ecological conditions. The review by Ah-King and Gowaty (2016) provides a large number of examples of such flexibility.

In summary, male investment seems to be a necessary, but not sufficient condition for the evolution of male mate choice. The role of differences in female quality and OSR seems to be important, too. And, potentially, all these mechanisms can interact to provide the basis for the expression of male mate choice.

5.6 Short summary

Differences in investment in gametes are the basis for the widely accepted sex roles, which predict that females benefit from being choosy, while males do not. In this chapter I discuss investments males make in either offspring or mates and ask whether any of those might be associated with male choice. The answer is complex, but it seems that male investment alone is not a very strong predictor of

male choosiness. Combined with sex ratio, however, investment seems to be quite important.

5.7 Further reading

The papers by Bateman (1948) and Trivers (1972) are classics in this context. A careful discussion of the complexity of male investment in female partners is provided by Sakaluk et al. (2019).

5.8 References

AH-KING, M. & GOWATY, P. A. 2016. A conceptual review of mate choice: stochastic demography, within-sex phenotypic plasticity, and individual flexibility. *Ecology and Evolution*, 6, 4607–42.

ALGER, I., HOOPER, P. L., COX, D., STIEGLITZ, J., & KAPLAN, H. S. 2020. Paternal provisioning results from ecological change. *Proceedings of the National Academy of Sciences of the United States of America*, 117, 10746–54.

ALVERGNE, A., FAURIE, C., & RAYMOND, M. 2009. Father–offspring resemblance predicts paternal investment in humans. *Animal Behaviour*, 78, 61–9.

ANDRADE, M. C. & KASUMOVIC, M. M. 2005. Terminal investment strategies and male mate choice: extreme tests of Bateman. *Integrative and Comparative Biology*, 45, 838–47.

ARMÚA DE REYES, C. & ESTÉVEZ, A. L. 2006. Predation on *Biomphalaria* sp. (Mollusca: Planorbidae) by three species of the genus *Belostoma* (Heteroptera: Belostomatidae). *Brazilian Journal of Biology*, 66, 1033–5.

BAKKER, T. C. M. & ROWLAND, W. J. 1995. Male mating preference in sticklebacks: effects of repeated testing and own attractiveness. *Behaviour*, 132, 935–49.

BALSHINE-EARN, S. 1996. Reproductive rates, operational sex ratios and mate choice in St. Peter's fish. *Behavioral Ecology and Sociobiology*, 39, 107–16.

BALSHINE-EARN, S. & EARN, D. J. 1998. On the evolutionary pathway of parental care in mouth-brooding cichlid fishes. *Proceedings of the Royal Society of London—Series B: Biological Sciences*, 265, 2217–22.

BATEMAN, A. J. 1948. Intra-sexual selection in *Drosophila*. *Heredity*, 2, 349–68.

BENUN SUTTON, F. & WILSON, A. B. 2019. Where are all the moms? External fertilization predicts the rise of male parental care in bony fishes. *Evolution*, 73, 2451–60.

BERGLUND, A., BISAZZA, A., & PILASTRO, A. 1996. Armaments and ornaments: an evolutionary explanation of traits of dual utility. *Biological Journal of the Linnean Society*, 58, 385–99.

BIRKHEAD, T. R., HOSKEN, D. J., & PITNICK, S. S. 2008. *Sperm Biology*. Amsterdam: Elsevier.

BLUMER, L. S. 1979. Male parental care in the bony fishes. *The Quarterly Review of Biology*, 54, 149–61.

BORGIA, G. 1993. The cost of display in the non-resource-based mating system of the satin bowerbird. *American Naturalist*, 141, 729–43.

BORGIA, G. 1995. Complex male display and female choice in the spotted bowerbird—specialized functions for different bower decorations. *Animal Behaviour*, 49, 1291–301.

BRADBURY, J. W. & VEHRENCAMP, S. 2011. *Principles of Animal Communication*. Sunderland, MA: Sinauer.

BRADBURY, J. W., VEHRENCAMP, S. L., & GIBSON, R. 1985. Leks and the unanimity of female choice. *In*: GREENWOOD, P. J., HARVEY, P. H., & SLATKIN, M. (eds.) *Evolution: Essays in Honour of John Maynard Smith*. Cambridge: Cambridge University Press, pp. 301–14.

CANDOLIN, U. & SALESTO, T. 2009. Does competition allow male mate choosiness in threespine sticklebacks? *American Naturalist*, 173, 273–7.

CARDOZO, G., DEVIGILI, A., ANTONELLI, P., & PILASTRO, A. 2020. Female sperm storage mediates post-copulatory costs and benefits of ejaculate anticipatory plasticity in the guppy. *Journal of Evolutionary Biology*.

CLUTTON-BROCK, T. H., HIRAIWA-HASEGAWA, M., & ROBERTSON, A. 1989. Mate choice on fallow deer leks. *Nature*, 340, 463–5.

CROWL, T. A. & ALEXANDER JR, J. E. 1989. Parental care and foraging ability in male water bugs (*Belostoma flumineum*). *Canadian Journal of Zoology*, 67, 513–15.

DAWKINS, R. 1982. *The Extended Phenotype*. Oxford: Oxford University Press.

DIAMOND, J. 1986. Biology of birds of paradise and bowerbirds. *Annual Review of Ecology and Systematics*, 17, 17–37.

EVANS, J. P., ROSENGRAVE, P., GASPARINI, C., & GEMMELL, N. J. 2013. Delineating the roles of males and females in sperm competition. *Proceedings of the Royal Society of London—Series B: Biological Sciences*, 280. https://doi.org/10.1098/rspb.2013.2047.

FIRMAN, R. C., GASPARINI, C., MANIER, M. K., & PIZZARI, T. 2017. Cryptic female choice: a general phenomenon. A reply to Eberhard. *Trends in Ecology & Evolution*, 32, 807–7.

GASPARINI, C., ANDREATTA, G., & PILASTRO, A. 2012. Ovarian fluid of receptive females enhances sperm velocity. *Naturwissenschaften*, 99, 417–20.

GEARY, D. C. 2015. Evolution of paternal investment. *In*: BUSS, D. M. (ed.) *The Handbook of Evolutionary Psychology*. Hoboken, NJ: John Wiley & Sons, pp. 1–18.

GOLDBERG, D. L., LANDY, J. A., TRAVIS, J., SPRINGER, M. S., & REZNICK, D. N. 2019. In love and war: the morphometric and phylogenetic basis of ornamentation, and the evolution of male display behavior, in the livebearer genus *Poecilia*. *Evolution*, 73, 360–77.

GOODWIN, N. B., BALSHINE-EARN, S., & REYNOLDS, J. D. 1998. Evolutionary transitions in parental care in cichlid fish. *Proceedings of the Royal Society of London—Series B: Biological Sciences*, 265, 2265–72.

GOWATY, P. A. & HUBBELL, S. P. 2009. Reproductive decisions under ecological constraints: it's about time. *Proceedings of the National Academy of Sciences of the United States of America*, 106, 10017–24.

GOWATY, P. A., KIM, Y.-K., & ANDERSON, W. W. 2012. No evidence of sexual selection in a repetition of Bateman's classic study of *Drosophila melanogaster*. *Proceedings of the National Academy of Sciences*, 109, 11740–5.

GWYNNE, D. T. 2008. Sexual conflict over nuptial gifts in insects. *Annual Review of Entomology*, 53, 83–101.

GWYNNE, D. & SIMMONS, L. 1990. Experimental reversal of courtship roles in an insect. *Nature*, 346, 172–4.

HÄRDLING, R. & KAITALA, A. 2001. Conflict of interest between sexes over cooperation: a supergame on egg carrying and mating in a coreid bug. *Behavioral Ecology*, 12, 659–65.

HÖGLUND, J. & ALATALO, R. V. 2014. *Leks*. Princeton, NJ: Princeton University Press.

ICHIKAWA, N. 1989. Breeding strategy of the male brooding water bug, *Diplonychus major* esaki (Heteroptera, Belostomatidae): is male back space limiting? *Journal of Ethology*, 7, 133–40.

JOHNSTONE, R. A. 1997. The tactics of mutual mate choice and competitive search. *Behavioral Ecology & Sociobiology*, 40, 51–9.

JOHNSTONE, R. A., REYNOLDS, J. D., & DEUTSCH, J. C. 1996. Mutual mate choice and sex differences in choosiness. *Evolution*, 50, 1382–91.

KIGHT, S. L. & KRUSE, K. C. 1992. Factors affecting the allocation of paternal care in waterbugs (*Belostoma* flumineum Say). *Behavioral Ecology and Sociobiology, 30,* 409–14.

KIGHT, S. L., TANNER, A. W., & COFFEY, G. L. 2011. Termination of parental care in male giant waterbugs, *Belostoma flumineum* Say (Heteroptera: Belostomatidae) is associated with breeding season, egg pad size, and presence of females. *Invertebrate Reproduction & Development,* 55, 197–204.

KOKKO, H. & JENNIONS, M. D. 2008. Parental investment, sexual selection and sex ratios. *Journal of Evolutionary Biology,* 21, 919–48.

KOKKO, H. & JOHNSTONE, R. A. 2002. Why is mutual mate choice not the norm? Operational sex ratios, sex roles and the evolution of sexually dimorphic and monomorphic signalling. *Proceedings of the Royal Society of London—Series B: Biological Sciences,* 357, 319–30.

KRAAK, S. B. M. & BAKKER, T. C. M. 1998. Mutual mate choice in sticklebacks: attractive males choose big females, which lay big eggs. *Animal Behaviour,* 56, 859–66.

KRUSE, K. C. 1990. Male backspace availability in the giant waterbug (*Belostoma flumineum* Say). *Behavioral Ecology and Sociobiology,* 26, 281–9.

LEHMANN, G. U. C. 2012. Weighing costs and benefits of mating in bushcrickets (Insecta: Orthoptera: Tettigoniidae), with an emphasis on nuptial gifts, protandry and mate density. *Frontiers in Zoology,* 19. https://doi.org/10.1186/1742-9994-9-19.

LEWIS, S. & SOUTH, A. 2012. The evolution of animal nuptial gifts. *In:* BROCKMANN, H. J., ROPER, T. J., NAGUIB, M., MITANI, J. C., & SIMMONS, L. W. (eds.) *Advances in the Study of Behavior, Vol 44.* Cambridge, MA: Academic Press.

LÜPOLD, S., MANIER, M. K., ALA-HONKOLA, O., BELOTE, J. M., & PITNICK, S. 2010. Male *Drosophila melanogaster* adjust ejaculate size based on female mating status, fecundity, and age. *Behavioral Ecology,* 22, 184–91.

LÜPOLD, S., MANIER, M. K., PUNIAMOORTHY, N., SCHOFF, C., STARMER, W. T., LUEPOLD, S. H. B., BELOTE, J. M., & PITNICK, S. 2016. How sexual selection can drive the evolution of costly sperm ornamentation. *Nature,* 533, 535-8.

LÜPOLD, S., REIL, J. B., MANIER, M. K., ZEENDER, V., BELOTE, J. M., & PITNICK, S. 2020. How female × male and male × male interactions influence competitive fertilization in *Drosophila melanogaster. Evolution Letters,* 4, 416–29.

MACARTNEY, E. L., CREAN, A. J., & BONDURIANSKY, R. 2018. Epigenetic paternal effects as costly, condition-dependent traits. *Heredity,* 121, 248–56.

MAGRIS, M., CARDOZO, G., SANTI, F., DEVIGILI, A., & PILASTRO, A. 2017. Artificial insemination unveils a first-male fertilization advantage in the guppy. *Animal Behaviour,* 131, 45–55.

MOBLEY, K. B. & JONES, A. G. 2013. Overcoming statistical bias to estimate genetic mating systems in open populations: a comparison of Bateman's principles between the sexes in a sex-role-reversed pipefish. *Evolution: International Journal of Organic Evolution,* 67, 646–60.

OBERZAUCHER, E. & GRAMMER, K. 2009. Ageing, mate preferences and sexuality: a mini-review. *Gerontology,* 55, 371–8.

OLIVEIRA, R. F., TABORSKY, M., & BROCKMANN, H. J. (eds.) 2008. *Alternative Reproductive Tactics.* Cambridge: Cambridge University Press.

PLATH, M., RIESCH, R., CULUMBER, Z., STREIT, B., & TOBLER, M. 2011. Giant water bug (*Belostoma* sp.) predation on a cave fish (*Poecilia mexicana*): effects of female body size and gestational state. *Evolutionary Ecology Research,* 13, 133–44.

RICHARDSON, J. & SMISETH, P. T. 2020. Maternity uncertainty in cobreeding beetles: females lay more and larger eggs and provide less care. *Behavioral Ecology,* 3, 641–50.

ROWLAND, W. J. 1982. Mate choice by male sticklebacks, *Gasterosteus aculeatus. Animal Behaviour,* 30, 1093–8.

ROWLAND, W. J. 1989. The effects of body size, aggression and nuptial coloration on competition for territories in male threespine sticklebacks, *Gasterosteus aculeatus. Animal Behaviour,* 37, 282–9.

ROWLAND, W. J., BAUBE, C. L., & HORAN, T. T. 1991. Signaling of sexual receptivity by pigmentation pattern in female sticklebacks. *Animal Behaviour,* 42, 243–9.

ROWLAND, W. J. & SEVENSTER, P. 1985. Sign stimuli in the threespine stickleback (*Gasterosteus aculeatus*): a re-examination and extension of some classic experiments. *Behaviour,* 93, 241–57.

SAKALUK, S. K., DUFFIELD, K. R., RAPKIN, J., SADD, B. M., & HUNT, J. 2019. The troublesome gift: the spermatophylax as a purveyor of sexual conflict and coercion in crickets. *Advances in the Study of Behavior,* 51, 1–30.

SCHLUPP, I. 2018. Male mate choice in livebearing fishes: an overview. *Current Zoology,* 64, 393–403.

SHELDON, B. & ELLEGREN, H. 1998. Paternal effort related to experimentally manipulated paternity of male collared flycatchers. *Proceedings of the Royal Society of London—Series B: Biological Sciences,* 265, 1737–42.

SILVA, K., VIEIRA, M. N., ALMADA, V. C., & MONTEIRO, N. M. 2010. Reversing sex role reversal: compete only when you must. *Animal Behaviour,* 79, 885–93.

SIROT, L. K. 2019. Modulation of seminal fluid molecules by males and females. *Current Opinion in Insect Science,* 35, 109–16.

SKUTCH, A. F. 1935. Helpers at the nest. *The Auk,* 52, 257–73.

SMITH, C., FLETCHER, D., WHRISKEY, F., & WOOTTON, R. 1995. A review of reproductive rates in sticklebacks in relation to parental expenditure and operational sex ratios. *Behaviour*, 132, 915–33.

SMITH, C. D. 2019. Mate selection for the female giant waterbug (*Belostoma flumineum* Say, 1832): for her, it's a coin toss or polyandry. *Aquatic Insects*, 40, 355–61.

SMITH, R. L. 1997. Evolution of paternal care in the giant water bugs (Heteroptera: Belostomatidae). *In*: CHOE, J. C. & CRESPI, B. J. (eds.) *The Evolution of Social Behavior in Insects and Arachnids*. Cambridge: Cambridge University Press, pp. 116–49.

SOGABE, A. & YANAGISAWA, Y. 2007. Sex-role reversal of a monogamous pipefish without higher potential reproductive rate in females. *Proceedings of the Royal Society of London—Series B: Biological Sciences*, 274, 2959–63.

SOUTH, A. & LEWIS, S. M. 2011. The influence of male ejaculate quantity on female fitness: a meta-analysis. *Biological Reviews*, 86, 299–309.

SPIKES, M. & Schlupp, I. 2021. Males can't afford to be choosy: male reproductive investment does not influence preference for female size in *Limia* (Poeciliidae). Behavioral Processes, 183, 104315.

TALLAMY, D. W. 2001. Evolution of exclusive paternal care in arthopods. *Annual Review of Entomology*, 46, 139–65.

THOMAS, L. K. & MANICA, A. 2005. Intrasexual competition and mate choice in assassin bugs with uniparental male and female care. *Animal Behaviour*, 69, 275–81.

THRASHER, P., REYES, E., & KLUG, H. 2015. Parental care and mate choice in the giant water bug *Belostoma lutarium*. *Ethology*, 121, 1018–29.

TINBERGEN, N. 1952. The curious behavior of the stickle-back. *Scientific American*, 187, 22–7.

TOBLER, M., SCHLUPP, I., & PLATH, M. 2007. Predation of a cave fish (*Poecilia mexicana*, Poeciliidae) by a giant water-bug (Belostoma, Belostomatidae) in a Mexican sulphur cave. *Ecological Entomology*, 32, 492–5.

TORRES-VILA, L. M. & JENNIONS, M. D. 2005. Male mating history and female fecundity in the Lepidoptera: do male virgins make better partners? *Behavioral Ecology and Sociobiology*, 57, 318–26.

TRIVERS, R. 1972. Parental investment and sexual selection. *In*: CAMPBELL, B. G. (ed.) Sexual Selection and the Descent of Man, 1871–1971. Chicago, IL: Aldine Publishing Company, pp. 136–79.

TRIVERS, R. 1974. Parent–offspring conflict. *American Zoologist*, 14, 249–64.

VALENCIA-AGUILAR, A., DE JESUS RODRIGUES, D., & PRADO, C. P. A. 2020. Male care status influences the risk-taking decisions in a glassfrog. *Behavioral Ecology and Sociobiology*, 74, 84.

VINCENT, A., AHNESJÖ, I., & BERGLUND, A. 1994. Operational sex ratios and behavioral sex differences in a pipefish population. *Behavioral Ecology and Sociobiology*, 34, 435–42.

VINCENT, A., AHNESJÖ, I., BERGLUND, A., & ROSENQVIST, G. 1992. Pipefishes and seahorses: are they all sex role reversed? *Trends in Ecology & Evolution*, 7, 237–41.

WIGGINS, D. A. & MORRIS, R. D. 1986. Criteria for female choice of mates: courtship feeding and paternal care in the common tern. *The American Naturalist*, 128, 126–9.

WOOLFENDEN, G. E. 1975. Florida scrub jay helpers at the nest. *The Auk*, 92, 1–15.

WOOTTON, R. J. 1984. *A Functional Biology of Sticklebacks*. Berkeley, CA: University of California Press.

WRIGHT, P. C. 1990. Patterns of paternal care in primates. *International Journal of Primatology*, 11, 89–102.

YONG, L., LEE, B. & MCKINNON, J. S. 2018. Variation in female aggression in 2 three-spined stickleback populations with female throat and spine coloration. *Current Biology*, 64, 345–50.

ZEH, D. W. & SMITH, R. L. 1985. Paternal investment by terrestrial arthropods. *American Zoologist*, 25, 785–805.

CHAPTER 6

Mechanism of Male Choice: Sex Ratios

6.1 Brief outline of the chapter

In this chapter I want to discuss the role of sex ratios in choosiness. So far, we have mostly reviewed intrinsic reasons for male choosiness to be expressed, such as male investment and female quality. However, sex ratios may also be important drivers of choosiness. Sex ratios are important in population biology and influence the evolution and structure of mating systems. Most important for the purpose of this book is that they can change quickly in time and space. Male and female choice are sensitive to such changes and can lead to situations where females are choosy when they are rare in a population but change to courtship and competition when males are rare. There are not many examples for this process, but there are likely some that have been overlooked. Interestingly, the majority of data on preferences are collected using binary choice tests, which almost always represent a 2:1 sex ratio. Furthermore, sex ratios do not take into account differences in mate quality, as all adult individuals are classified as either male or female without making any further distinction.

6.2 Introduction

Sex ratios are important in ecology and evolution. Typically, they are balanced and even at conception (primary sex ratio) and birth (secondary sex ratio), but they can shift to either male- or female-biased as cohorts of individuals sexually mature. The balanced 1:1 sex ratio is often referred to as Fisher's principle and represents an evolutionarily stable strategy (ESS). If producing male and female offspring has the same cost, deviations from a 1:1 sex ratio are not evolutionarily stable (Hamilton, 1967), an example of frequency-dependent selection.

At later points in life, however, sex ratios are often biased owing to a number of factors that selectively affect either males or females (Kvarnemo and Ahnesjö, 1996; Székely et al., 2014). To characterize the part of a population that can potentially reproduce, two measures are widely used. One is the adult sex ratio (ASR) (Mayr, 1939), the number of individuals of one sex relative to the number of the other sex. The second measure is the operational sex ratio (OSR) (Emlen and Oring, 1977), reflecting the numbers of each sex ready to mate. This concept is intertwined with the concept of the potential reproductive rate (PRR), which is defined as the offspring production relative to time for each sex assuming unconstrained mate availability. This rate often differs between males and females and can be used to predict mating competition (Ahnesjö et al., 2001). Although clearly related, OSR and ASR measure different aspects of a population and do not provide the same information.

Intuitively, one can relate to the argument that whichever sex is rarer can be more selective in mate choice, simply because there are more candidates to choose from. Indeed, limitation of access to mates is an important element of sexual selection. Arguably, sex ratios alone can drive choosiness, but a more complete view includes sex ratios and investment in gametes. Both the sex ratio and investment in gametes are used to measure and predict directionality and strength of sexual selection (Kvarnemo and Ahnesjö, 1996). They can be especially powerful when combined (Kokko et al., 2012). A third factor influencing the directionality and strength of male mate choice (or mate choice in general) is the quality of the partner, or for male mate choice, female quality (Chapter 4). These factors can interact in multiple ways (Figure 6.1).

Male Choice, Female Competition, and Female Ornaments in Sexual Selection. Ingo Schlupp, Oxford University Press (2021). © Ingo Schlupp (2021).
DOI: 10.1093/oso/9780198818946.003.0006

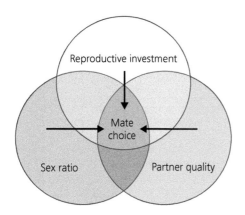

Figure 6.1 Possible interactions of three variables influencing mate choice, partner quality, sex ratio, and reproductive investment.

6.3 OSR

One common way to characterize the sex ratio in a population is to calculate how many individuals of one sex are ready to reproduce (or mate) relative to the other sex (Emlen and Oring, 1977; Kvarnemo and Ahnesjö, 1996) at a given time and place. Defined as the ratio of males ready to mate relative to females ready to mate, often the OSR is male-biased. The stronger the bias, the more likely males will compete for females. The OSR is also thought to be important in the evolution and strength of female choice: the more males, the choosier a female can be, as was shown for example in guppies (*Poecilia reticulata*) (Jirotkul, 1999). In other words, if the sex ratio is male-biased, males should compete with males, but if it is female-biased, females should compete for males. As I will argue below, however, this approach ignores that certain, high-quality mating partners may be rare, even when partners in general are abundant. This could mean that some choosers have to settle for lower-quality mates. A highly biased sex ratio can also come with costs (Magurran and Nowak, 1991): if males, for example, are very abundant, competition may become fierce and lead to increased rates of injury, and—maybe more importantly—males may benefit from undermining and thwarting female choice. This form of sexual conflict can lead to males showing harassment (Magurran and Seghers, 1994; Schlupp et al., 2001; Plath et al., 2007) and alternative mating strategies (Oliveira et al., 2008). An abundance of

males has also been suggested as the backdrop for violence in humans ("more men, more violence"), but Schacht et al. (2014) argue that this explanation is too simplistic. In their review they also point to the usefulness of ASR as an alternative way of understanding sexual selection in humans.

We have already encountered a number of examples from the literature that make reference to the role of OSR in male mate choice in previous chapters. ASR, on the other hand, has been used less to understand mate choice (Grant and Grant, 2019), but has been suggested to be generally important (Székely et al., 2014; Kappeler, 2017) (but see Jennions and Fromhage, 2017) and, for example, drive sex-role reversal in shorebirds (Liker et al., 2013). The OSR, on the other hand, is heavily influenced by the time individuals are not able to reproduce after a mating event. For various reasons, after mating, individuals of many species are unable to engage in another mating for some time, taking them out of the mating pool. This timeout period is often associated with caring for offspring because typically an individual that is caring for offspring is unable to engage in additional matings. An example of this would be mouthbrooding in female cichlids, where females are not only out of the pool of breeding females, but will not even eat while they provide for the young (Schürch and Taborsky, 2005). Consequently, even with a balanced ASR, the OSR can be strongly biased. In a bush cricket from Australia, *Kawanaphila mirla*, the timeout period for females was six times longer than that for males (Lehmann and Lehmann, 2007), making the OSR male-biased, even when the ASR was even. In this group of crickets, males produce very large spermatophores and the males are only able to produce these large spermatophores under good environmental conditions. As the environmental conditions worsen, ASR and OSR will approach even values and the OSR can then flip, which has been shown in a related species, *Kawanaphila nartee* (Gwynne and Simmons, 1990). In another group, mammals, the females are out of the pool of individuals ready to mate owing to pregnancy and lactation, which is relatively long and here too the OSR becomes male-biased, while the ASR can remain even (Kappeler, 2017). Of course, males can also be away from the pool of individuals ready to mate and take a timeout. Calling, for example can be energetically demanding for males in a frog chorus, and males may have to leave a chorus to replenish their energy (Runkle et al., 1994). This will then alter the OSR locally and temporally and make it less male-biased. The timeout period can be short or long relative to life expectancy. In men, the refractory period after an ejaculation ranges from a few minutes to an hour (Jones and Lopez, 2013), which is negligible given the life span. The same might be true for most mammals. In species, however, where males show parental care, they can be removed from the pool of reproducing individuals for a long time. This intersects with ideas relevant to male investment: as males invest time and energy in their offspring, they face tradeoffs and may miss opportunities to mate. This in turn could favor males that are choosier.

Quite often females are the rarer sex. And often females also invest more in offspring as compared with males. Consequently, the choosiness predicted based on sex ratio (OSR) and investment in gametes (Bateman gradient) will often show the same directionality and both may be used to explain the prevalence of female choice. However, as Kokko et al. (2012) argue, this does not always have to be the case. If the OSR is balanced, one would predict relatively weak sexual selection (Owens and Thompson, 1994). However, the sex ratio is only one of several factors to influence choosiness.

There are two aspects of OSR and ASR that are—in my view—not well understood. The first one has to do with how we generate data on mate choice; the second has to do with partner quality. OSR and ASR can fluctuate considerably in nature and these fluctuations can be important to the evolution or expression of male mate choice. Yet, the common method to collect preference of choice data does not capture the variability we observe in OSRs. We often, maybe typically even, measure mate preferences and choice in an experimental setting that involves one chooser and two stimuli. This type of choice test, a binary choice test, provides data on mating preferences in many different contexts and using most sensory channels (Rosenthal, 2017), ranging from visual choice tests to acoustical and chemical information. In a way, binary choice tests are the workhorse of mate choice research (Figure 6.2). One consequence of this is, however,

that almost all of the experimental data on female choice are collected under a 2:1 sex ratio. After all, in a typical binary choice test, one chooser selects among two courters. Creating other sex ratios in choice experiments might be difficult, because experimentally it might then be difficult to take attraction to groups into account. But using the same sex ratio for most experiments might also be problematic. More generally, experimental manipulations of sex ratios to study its effects on mate choice are relatively rare. One example is provided by a study on Jamaican field crickets (*Gryllus assimilis*), which created even, male-biased, and female-biased sex ratios and tested both males and females under the three different sex ratios (Villarreal et al., 2018). This study partly supported the switch point theorem (Gowaty and Hubbell, 2009; Ah-King and Gowaty, 2016). This idea explicitly suggests that mating decisions by individuals should be flexible based on ecology (Chapter 5). Another example is a study by Hayes using robotic crabs (Hayes et al., 2016). Another example is a study in domestic pigeons, *Columba livia*, where a male-biased ASR was created during part of the breeding season. During that time divorces went up and more clutches failed (Marchesan, 2002). This is in agreement with general predictions that a male-biased sex ratio can lead to male behaviors such as harassment of females or increased aggression between males with adverse effects on the females.

This dynamic nature of OSR in mate choice has been identified as an issue before (Hayes et al., 2016), but is still mostly not incorporated into the design of choice experiments. As Hayes et al. (2016) elaborate, the situation can actually be even more complex. Mating itself can influence the local OSR: if an individual picks a mating partner, this mating

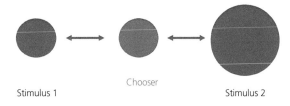

Stimulus 1   Chooser   Stimulus 2

Figure 6.2 Cartoon of a binary choice test set-up. The chooser can pick between two stimuli that differ in the trait of interest, in this hypothetical example, size.

partner is then removed from the mating pool for some time, thus altering the OSR. Just like males dropping out to replenish their resources, this process can change the sex ratio quickly, albeit temporarily. Hayes et al. (2016) use a study on tungara frogs (Ryan, 1985) to illustrate this point: 20 females encounter 100 calling males. If the females arrive asynchronously at the breeding site, the first female that selects a male and removes him from the mating pool for that night alters the sex ratio to 19 females to 99 males, and so on. For the 20th female the sex ratio is 1:81. To be clear, this scenario assumes, of course, that the males are removed from the pool for at least as long as the 20 females are choosing. Such a pattern is, however, fairly common. It is especially the case in explosive breeding species. For example, in common toads from Europe (*Bufo bufo*), males arrive earlier than females at any given breeding site. A female that arrives is typically in amplexus with one male for the remainder of the breeding season, which often lasts for several more days. Thus, early on the ratio of males to females may be less biased than later in the season. It seems that males respond to this in an adaptive way and, early in the season when male density is low, they court females (Höglund, 1988). They change their behavior later to scramble competition when male densities are much higher and the OSR is more male-biased (Davies and Halliday, 1978).

The second interesting issue with using OSR to characterize mating systems is that OSR looks strictly at numbers. To a researcher measuring OSR, all males and females are equal. Yet, clearly females and males also differ in quality. Consequently, removing individuals of different quality may influence the mating pool differently. If the most "attractive" mates are removed from the mating pool first, the remaining individuals may adjust their preferences. For example, if the top-ranked males are out of the pool for a relatively short time—for example, because they are sperm depleted—it might be preferable to females to wait for the top males to return, as compared with mating with other males. Even if males in general are abundant, this is not true for high-quality males. They are—by definition—rare. From a male point of view, this creates an interesting scenario where, even with a male-biased OSR, some males may exercise choice, whereas others

may not. In many other areas too, a more individualized approach to mate choice would be useful.

Overall, however, what makes OSR as a concept useful and attractive to me is the fact that the OSR can change quickly and reflects the ecology of the given species at a local spatial scale and on a relatively short temporal scale. Spatial and temporal variation in OSR or ASR is often detected in long-term studies (Grant and Grant, 2019), and has also been reported for humans (Stone et al., 2007).

Generally, OSR can be used to make predictions on competition and courtship behavior (de Jong et al., 2012; Myhre et al., 2012). If the sex ratio predicts which sex is more likely to choose, changes in sex ratio should predict changes in choosiness. Hence, for situations where the OSR can significantly change or flip over the course of a reproductive season or a limited area, flexible behavioral strategies are predicted (Gowaty and Hubbell, 2009; Villarreal et al., 2018). We have already briefly encountered an example for that in Chapter 1: male and female two-spotted gobies (*Gobiusculus flavescens*), a widespread European species with a coastal distribution, can adjust their mating behavior based on seasonal changes in OSR (Forsgren et al., 2004; Amundsen, 2018). These gobies have a short breeding season in the northern part of their distribution and after about half of the season the OSR changes from male-biased to female-biased. With the change in OSR male and female behavior also switches: females start the season being choosy, but then as the OSR changes, they start courting males as those become rare. And males start out courting and competing but switch over to being choosy as they become increasingly rare. In this system females also have ornaments, as discussed in Chapter 7. What is especially important about this example is that a relatively short-term, yet predictable, temporal ecological change is altering the dynamics of mate choice. Importantly, behaviors typically associated with one sex role are expressed by both sexes, indicating again how flexible sexual behavior can be and that sex roles may not be a good descriptor in these cases.

On a spatial scale there can also be population variation for OSR. This means that under this scenario, the ability of males to choose may be expressed in some individuals or populations, but not in others. In a widespread coastal fish from the Mediterranean, the peacock blenny (*Salaria pavo*), males usually compete for females when breeding opportunities are abundant and the sex ratio is male-biased, but in the Ria Formosa mudflat in the Portuguese Algarve, where breeding sites are rare, the sex ratio changes and few males are ready to mate (Almada et al., 1994). As predicted based on ecology and OSR, males of that population are choosy and females compete for males (Almada et al., 1995). Peacock blennies nest in natural rock crevices throughout their Mediterranean range, but those are naturally absent in the soft mud flats of a stretch of coastline in Portugal, the Algarve. However, in the Algarve clam culturists use cored building bricks to delimit their fields (Almada et al., 1994). This provides a substitute for the rock crevices males use elsewhere throughout their range. Nonetheless, in the Algarve usable nesting sites for males remain rare and become a limiting resource. This leads to strong competition for nesting sites among males, but also changes the OSR to female-biased and allows males to be choosy. Thus, it is a recent human change to the habitat—ecological engineering—that allows peacock blennies to breed there. The breeding behavior, however, apparently adjusted very quickly from female choice to male choice and from male courtship to female courtship. This suggests again that male and female strategies are more flexible than often assumed. In another potential example, in the Neotropical frog, *Thoropa taophora*, a somewhat similar ecological scenario was reported. In this species males also monopolize rare breeding sites and show site fidelity (de Sá et al., 2020). Whether males are choosy, however, is not clear yet.

Yet another example of the important role of the sex ratio—this time related to male investment—is provided by an Australian bush cricket. In *Kawanaphila nartee*, males provide a costly, energy-rich spermatophore to females (Gwynne et al., 1998; Gwynne and Simmons, 1990). Under conditions where males find abundant food, they have no problem providing a spermatophore to females and the OSR is male-biased. In harsher conditions, however, the OSR flips because fewer males are able to forage well enough to obtain sufficient energy to build a spermatophore (Simmons, 1992). As a con-

sequence, under low-food conditions, males are choosy, and females compete for the few males that can provide a spermatophore (Simmons and Bailey, 1990). Interestingly, as conditions become more adverse not only male quality shows high variance, but female quality too. This in turn interacts with male mate choice (Kvarnemo and Simmons, 1999). Furthermore, this example highlights the intersection of male investment (Chapter 5): males become limiting for females because of their significant investment in their offspring via a large spermatophore.

Overall, these examples show how changes in the OSR can dynamically lead to adjustments in "sex roles," based on the underlying ecology and consequently lead to male mate choice, whether on a species level or much more flexible on a population level. It is noteworthy again that male choice is expressed immediately when the conditions are favorable. However, in general, surprisingly few examples are known where OSR-based male mate choice is found (Chapter 2), even when we include the sex-role reversed species. Some examples may be overlooked, but overall it seems that being rare, especially locally and for a short time, is not enough to allow for the evolution of male mate choice.

6.3.1 What influences the OSR?

The OSR is influenced by many factors and in turn influences many others. Considering these factors may not seem directly important to understanding the connection between OSR and male mate choice, but because a female-biased OSR may select for male mate choice, knowing how biases in OSR arise can be important. For example, combined with information on the PRR, the OSR allows predictions of the direction and intensity of sexual selection (Kvarnemo and Ahnesjö, 1996). Differential mortality of the sexes is one of the many factors that influence the OSR. In sailfin mollies (*Poecilia latipinna*), a livebearing fish from Florida, sex-specific predation was detected against a complex background of mortality not related to predators (Trexler et al., 1992, 1994). In these fishes females are on average bigger than males, and are preferred by some predators, while others seem to prefer males (Trexler et al., 1994). Generally, however, males, with their showy ornamental traits and behaviors, are generally more likely to attract predators and be eaten (Sommer, 2000; Boukal et al., 2008; Han et al., 2015). Even in a cave environment, fish predation by a giant water bug, *Belostoma* sp., can be male-biased (Tobler et al., 2008). This is surprising as males are on average smaller in the prey, a cave-dwelling form of the livebearing fish *Poecilia mexicana*. Somewhat surprisingly, this example of male-biased predation is not associated with risky male behavior. Finally, humans can be an agent of sex-specific mortality by targeting predominantly males in hunting and harvesting (Clark and Tait, 1982).

Despite male specific predation, both ASR and OSR often count more males than females. Also in humans mortality is male-biased, in this case probably owing to risky behaviors shown by adolescent men, leading to accidents, homicide, and suicides (Mohler and Earls, 2001). Some of these behaviors may be meant to impress potential mates (Pianka, 2000). However, men at any age are more likely to die than women (Case and Paxson, 2005). Therefore, it seems counterintuitive that sex ratios are often male-biased although males are more likely to die. However, there are several mechanisms that can account for this, including earlier maturation of males. Aside from risky behavior associated with male courtship and mating, ornaments *per se* are thought to make males more vulnerable to predation. Ornamentation in females might also increase predation (Chapter 7).

In addition, the sexes may show different patterns in migration and dispersal. Sex-specific migration is very common in birds, often with the males migrating earlier (Briedis et al., 2019). Interestingly, sex-specific migration has also been suggested as the cause for male-biased sex ratios in rural East Germany, where women were more likely to leave after the German unification in 1990 (Leibert, 2016). Sex-specific dispersal is also common, with many driving factors behind it (Li and Kokko, 2019). For humans (*Homo sapiens*) variation in ASR and dispersal of the globally rarer sex has been suggested as important in the evolution of sociality (Kramer et al., 2017).

As I said above, differences in the time to reaching sexual maturity may also impact the OSR. If one sex takes longer to develop, an overabundance of the other sex is predicted. With males often reaching maturity earlier, again a male-biased sex ratio is predicted. It should be noted, however, that in many species maturation takes a long time; in humans this is a multi-year process (Jones and Lopez, 2013). Furthermore, not all individuals of one sex have to mature at the same time. In some livebearing fishes of the family Poeciliidae, for example, some males mature smaller (and younger), while others mature larger (and older) (Ryan et al., 1990; Schartl et al., 1993; Erbelding-Denk et al., 1994; Ptacek and Travis, 1996).

Finally, the sexes often differ in how long a reproductive event takes them out of the population before they can breed again. The outcome of this is heavily influenced by which sex is providing care for the offspring. In the case of female mammals, for example, this would include at a minimum the time of pregnancy and lactation. If males are the sole providers of parental care, the same reasoning applies to them, but in birds and mammals, they may seek extra reproductive events on the side. This is not possible for females as they have to wait until they ovulate again in many species; the sex ratio can also be influenced by immigration or emigration of individuals. On a very short timeline, it can change quickly in fission–fusion societies.

Alternatively, after a loss of offspring, for example owing to clutch loss, females can return quickly to the pool of reproductive individuals. This mechanism is thought to be behind infant killing in some species such as lions and several other species (Pusey and Packer, 1994). Males kill the offspring of other males to induce estrus in the females and sire their own offspring. Thus, effectively males are shortening the timeout period of the females to the males' benefit.

Another important aspect in this context is how synchronized the sexes are in their reproduction. If both sexes are ready to mate at the same time, the OSR will be relatively balanced, but if individuals of one sex are asynchronous in their availability, an unbalanced OSR will result (Kappeler, 2017). One consequence of a more biased sex ratio is usually intense competition among the more common sex, sometimes with negative feedback for the less common sex: if for example males are more common, they may evolve behaviors not only to compete with other males, but also to thwart female choice, such as sexual harassment.

The OSR may also be impacted by climate change, as was suggested for fishes (Geffroy and Wedekind, 2020). This is especially pertinent for species with temperature-dependent sex determination (TSD), like many turtles. Theoretically, rising temperatures should drive the sex ratios in the direction of the sex that is produced at higher temperatures. This was found for sea turtles, but the temperature at the nesting site would have to rise by 5°C to lead to population extinction (Hays et al., 2017).

6.3.2 What is influenced by the OSR?

Sex ratios in general have important effects on populations, but most of them are not the focus of this book. They are viewed as an organizing principle for mating systems (Shuster and Wade, 2003). One example shows how sex ratios, both OSR and ASR, can be important for the social environment, which I have already identified as critical: a study using mice (*Mus musculus*), shows that high male density can trigger adaptive female stress, which leads females to upregulate the number of daughters (Firman, 2020). These daughters will then likely experience an environment with many males to choose from. Generally, strongly male-biased sex ratios are likely to increase sexual conflict, including sexual harassment and male mating strategies that attempt to thwart female choice. In addition, the presence of many males in the population can lead to increased male competition, and alternative mating behaviors (Oliveira et al., 2008), including polyandry. On the population level, male-biased sex ratios also increase the asymmetry in reproductive success as fewer males actually reproduce. However, in a study using the European common lizard, *Lacerta vivipara*, a manipulation of the ASR revealed that sexual selection on male size was not predicted by ASR, suggesting that the costs of breeding were important, but not OSR (Fitze and Le Galliard, 2008).

Another response to changes in OSR can be adjusting parental investment. For example, an Australian species of lizard, *Amphibolurus muricatus*, was found to respond to biased OSRs by modifying their offspring sex ratio (Warner and Shine,

2007). The direction of the skew was more in agreement with a scenario in which this manipulation is driven by natural selection, not sexual selection. Under sexual selection and a male-biased sex ratio, it would be beneficial to produce more daughters, assuming they would have many males to choose from, but Warner and Shine (2007) found that more males were produced. This can be explained by assuming that males have a survival benefit and it pays to produce more sons. In a cichlid fish, *Sarotherodon galilaeus*, a relative of the widespread African genus *Tilapia* from Lake Kinneret in Israel, both sexes can show mouthbrooding in the same population. Brooding males were more likely to abandon their brood in an experiment that created a female-biased sex ratio, probably owing to more opportunities to remate. Females showed a similar pattern and were more likely to desert a brood under a male-biased sex ratio (Balshine-Earn and Earn, 1998). In a pipefish from Europe, *Syngnathus typhle*, choosiness of males was also dependent on OSR: in a female-biased situation, males preferred larger females in a binary choice test (Berglund, 1994). This is the normal situation for this sex-role reversed fish. In an experiment, however, males were given the impression that the sex ratio was male-biased, which led to a loss of the preference and random mating (Berglund, 1994). This is an especially important finding because this also indicates that choosers can adjust their choosiness on a local level and at short notice, including in sex-role reversed species.

Overall, and much simplified, a male-biased OSR is not conducive to the evolution of male mate choice in general, although a few high-quality males may still be choosy.

6.4 ASR

One problem with trying to capture the direction and strength of sexual selection in just one number is that nonbreeding individuals can play important roles without breeding. Some may serve as helpers or play a role in the social environment in which mate choice happens. Additional, nonbreeding males may for example influence mating by sexually harassing females, a very common behavior.

The ASR has been suggested as driving the evolution of mating systems in birds (Donald, 2007) and beyond also in other taxa (Székely et al., 2014), including the evolution of sex-role reversed mating systems (Liker et al., 2013). The association of sex roles and ASR is strong in shorebirds and predicts whether or not males should provide for offspring or not. This makes ASR a viable alternative to other explanations for sex-role reversal (Liker et al., 2013) that are based on mating effort and male investment. Furthermore, ASR has been found to have a role in mate choice in Darwin's finches (Grant and Grant, 2019). In these birds from the Galapagos Islands, harsh environmental conditions and sex-specific mortality can lead to a male-biased sex ratio, providing females the opportunity to choose and also secure multiple males and show polyandry. In addition the number of extra-pair matings and matings with another species—and hence hybridization—increased (Grant and Grant, 2019). However, the importance of ASR is not universally accepted (Jennions and Fromhage, 2017).

Overall, maybe the biggest practical concern with OSR and ASR is that they are not easily quantified. In many species one or both sexes are cryptic, making reliable counts difficult. If only one sex is conspicuous it is possible that some bias in counting may occur. In many species, it can be hard to determine which individuals are at least potentially reproducing, especially if some individuals are nonbreeding and may be missed in counts easily. Nonetheless, sex ratios are important for our understanding of the ecology and evolution of mate choice.

6.5 Short summary

Most importantly from my point of view, the OSR is fairly dynamic and can change quickly and locally. This makes it likely that choosiness based on OSR is not fixed but can be modified rather quickly.

In essence, we are saying that if males are rare, they might benefit from being choosy. What makes this argument different from the others we looked at before is that sex ratio can be a very dynamic population parameter. This suggests that preferences based on the sex ratio should not be fixed strategies, but flexible.

6.6 Further reading

A general and accessible overview of OSR is provided by Kvarnemo and Ahnesjö (1996) and a good introduction to ASR by Székely et al. (2014). An interesting review on sex ratios was published by Jennions and Fromhage (2017). A review of flexibility of mate choice based on ecological conditions was published by Ah-King and Gowaty (2016).

6.7 References

AH-KING, M. & GOWATY, P. A. 2016. A conceptual review of mate choice: stochastic demography, within-sex phenotypic plasticity, and individual flexibility. *Ecology and Evolution*, 6, 4607–42.

AHNESJÖ, I., KVARNEMO, C., & MERILAITA, S. 2001. Using potential reproductive rates to predict mating competition among individuals qualified to mate. *Behavioral Ecology*, 12, 397–401.

ALMADA, V. C., GONÇALVES, E. J., OLIVEIRA, R. F., & SANTOS, A. J. 1995. Courting females: ecological constraints affect sex roles in a natural population of the blenniid fish *Salaria pavo*. *Animal Behaviour*, 1995, 1125–7.

ALMADA, V. C., GONÇALVES, E. J., SANTOS, A. J., & BAPTISTA, C. 1994. Breeding ecology and nest aggregations in a population of *Salaria pavo* (Pisces: Blenniidae) in an area where nest sites are very scarce. *Journal of Fish Biology*, 45, 819–30.

AMUNDSEN, T. 2018. Sex roles and sexual selection: lessons from a dynamic model system. *Current Zoology*, 64, 363–92.

BALSHINE-EARN, S. & EARN, D. J. 1998. On the evolutionary pathway of parental care in mouth-brooding cichlid fishes. *Proceedings of the Royal Society of London—Series B: Biological Sciences*, 265, 2217–22.

BERGLUND, A. 1994. The operational sex ratio influences choosiness in a pipefish. *Behavioral Ecology*, 5, 254–8.

BOUKAL, D. S., BEREC, L., & KŘIVAN, V. 2008. Does sex-selective predation stabilize or destabilize predator-prey dynamics? *PLoS ONE*, 3, e2687.

BRIEDIS, M., BAUER, S., ADAMÍK, P., ALVES, J. A., COSTA, J. S., EMMENEGGER, T., GUSTAFSSON, L., KOLEČEK, J., LIECHTI, F., & MEIER, C. M. 2019. A full annual perspective on sex-biased migration timing in long-distance migratory birds. *Proceedings of the Royal Society of London—Series B: Biological Sciences*, 286, 20182821.

CASE, A. & PAXSON, C. 2005. Sex differences in morbidity and mortality. *Demography*, 42, 189–214.

CLARK, C. W. & TAIT, D. E. 1982. Sex-selective harvesting of wildlife populations. *Ecological Modelling*, 14, 251–60.

DAVIES, N. B. & HALLIDAY, T. R. 1978. Deep croaks and fighting assessment in toads *Bufo bufo*. *Nature*, 274, 683.

DE JONG, K., FORSGREN, E., SANDVIK, H., & AMUNDSEN, T. 2012. Measuring mating competition correctly: available evidence supports operational sex ratio theory. *Behavioral Ecology*, 23, 1170–7.

DE SÁ, F. P., CONSOLMAGNO, R. C., MURALIDHAR, P., BRASILEIRO, C. A., ZAMUDIO, K. R., & HADDAD, C. F. B. 2020. Unexpected reproductive fidelity in a polygynous frog. *Science Advances*, 6, eaay1539.

DONALD, P. F. 2007. Adult sex ratios in wild bird populations. *Ibis*, 149, 671–92.

EMLEN, S. T. & ORING, L. W. 1977. Ecology, sexual selection, and the evolution of mating systems. *Science*, 197, 215–23.

ERBELDING-DENK, C., SCHRODER, J. H., SCHARTL, M., NANDA, I., SCHMID, M., & EPPLEN, J. T. 1994. Male polymorphism in *Limia perugiae* (Pisces: Poeciliidae). *Behavioral Genetics*, 24, 95–101.

FIRMAN, R. C. 2020. Exposure to high male density causes maternal stress and female-biased sex ratios in a mammal. *Proceedings of the Royal Society of London—Series B: Biological Sciences*, 287, 20192909.

FITZE, P. S. & LE GALLIARD, J. F. 2008. Operational sex ratio, sexual conflict and the intensity of sexual selection. *Ecology Letters*, 11, 432–9.

FORSGREN, E., AMUNDSEN, T., BORG, Å. A., & BJELVENMARK, J. 2004. Unusually dynamic sex roles in a fish. *Nature*, 429, 551.

GEFFROY, B. & WEDEKIND, C. 2020. Effects of global warming on sex ratios in fishes. *Journal of Fish Biology*. https://doi.org/10.1111/jfb.14429

GOWATY, P. A. & HUBBELL, S. P. 2009. Reproductive decisions under ecological constraints: it's about time. *Proceedings of the National Academy of Sciences*, 106, 10017–24.

GRANT, P. R. & GRANT, B. R. 2019. Adult sex ratio influences mate choice in Darwin's finches. *Proceedings of the National Academy of Sciences*, 116, 12373–82.

GWYNNE, D. & SIMMONS, L. 1990. Experimental reversal of courtship roles in an insect. *Nature*, 346, 172–4.

GWYNNE, D. T., BAILEY, W. J., & ANNELLS, A. 1998. The sex in short supply for matings varies over small spatial scales in a katydid (*Kawanaphila nartee*, Orthoptera: Tettigoniidae). *Behavioral Ecology and Sociobiology*, 42, 157–62.

HAMILTON, W. D. 1967. Extraordinary sex ratios. *Science*, 156, 477–88.

HAN, C. S., JABLONSKI, P. G., & BROOKS, R. C. 2015. Intimidating courtship and sex differences in predation risk lead to sex-specific behavioural syndromes. *Animal Behaviour*, 109, 177–85.

HAYES, C. L., CALLANDER, S., BOOKSMYTHE, I., JENNIONS, M. D., & BACKWELL, P. R. Y. 2016. Mate choice and the operational sex ratio: an experimental test with robotic crabs. *Journal of Evolutionary Biology*, 29, 1455–1461.

HAYS, G. C., MAZARIS, A. D., SCHOFIELD, G., & LALOË, J.-O. 2017. Population viability at extreme sex-ratio skews produced by temperature-dependent sex determination. *Proceedings of the Royal Society of London—Series B: Biological Sciences*, 284, 20162576.

HÖGLUND, J. 1988. Chorusing behaviour, a density dependent alternative strategy in male common toads (*Bufo bufo*). *Ethology*, 79, 324–32.

JENNIONS, M. D. & FROMHAGE, L. 2017. Not all sex ratios are equal: the Fisher condition, parental care and sexual selection. *Philosophical Transactions of the Royal Society B—Biological Sciences*, 372, 20160312.

JIROTKUL, M. 1999. Operational sex ratio influences female preference and male–male competition in guppies. *Animal Behaviour*, 58, 287–94.

JONES, R. E. & LOPEZ, K. H. 2013. *Human Reproductive Biology*. London: Academic Press.

KAPPELER, P. M. 2017. Sex roles and adult sex ratios: insights from mammalian biology and consequences for primate behaviour. *Philosophical Transactions of the Royal Society B—Biological Sciences*, 372, 20160321.

KOKKO, H., KLUG, H., & JENNIONS, M. D. 2012. Unifying cornerstones of sexual selection: operational sex ratio, Bateman gradient and the scope for competitive investment. *Ecology Letters*, 15, 1340–51.

KRAMER, K. L., SCHACHT, R., & BELL, A. 2017. Adult sex ratios and partner scarcity among hunter–gatherers: implications for dispersal patterns and the evolution of human sociality. *Philosophical Transactions of the Royal Society B—Biological Sciences*, 372, 20160316.

KVARNEMO, C. & AHNESJÖ, I. 1996. The dynamics of operational sex ratios and competition for mates. *Trends in Ecology & Evolution*, 11, 404–8.

KVARNEMO, C. & SIMMONS, L. W. 1999. Variance in female quality, operational sex ratio and male mate choice in a bushcricket. *Behavioral Ecology and Sociobiology*, 45, 245–52.

LEHMANN, G. U. & LEHMANN, A. W. 2007. Sex differences in "time out" from reproductive activity and sexual selection in male bushcrickets (Orthoptera: Zaprochilinae: *Kawanaphila mirla*). *Journal of Insect Behavior*, 20, 215–27.

LEIBERT, T. 2016. She leaves, he stays? Sex-selective migration in rural East Germany. *Journal of Rural Studies*, 43, 267–79.

LI, X. Y. & KOKKO, H. 2019. Sex-biased dispersal: a review of the theory. *Biological Reviews*, 94, 721–36.

LIKER, A., FRECKLETON, R. P., & SZEKELY, T. 2013. The evolution of sex roles in birds is related to adult sex ratio. *Nature Communications*, 4, 1–6.

MAGURRAN, A. E. & NOWAK, M. A. 1991. Another battle of the sexes: the consequences of sexual asymmetry in mating costs and predation risk in the guppy, *Poecilia reticulata*. *Proceedings of the Royal Society of London— Series B: Biological Sciences*, 246, 31–8.

MAGURRAN, A. E. & SEGHERS, B. H. 1994. A cost of sexual harassment in the guppy, *Poecilia reticulata*. *Proceedings of the Royal Society of London—Series B: Biological Sciences*, 258, 89–92.

MARCHESAN, M. 2002. Operational sex ratio and breeding strategies in the feral pigeon *Columba livia*. *Ardea*, 90, 249–57.

MAYR, E. 1939. The sex ratio in wild birds. *The American Naturalist*, 73, 156–79.

MOHLER, B. & EARLS, F. 2001. Trends in adolescent suicide: misclassification bias? *American Journal of Public Health*, 91, 150.

MYHRE, L. C., DE JONG, K., FORSGREN, E., & AMUNDSEN, T. 2012. Sex roles and mutual mate choice matter during mate sampling. *American Naturalist*, 179, 741–55.

OLIVEIRA, R. F., TABORSKY, M., & BROCKMANN, H. J. (eds.) 2008. *Alternative Reproductive Tactics*. Cambridge: Cambridge University Press.

OWENS, I. P. F. & THOMPSON, D. B. A. 1994. Sex differences, sex ratios and sex roles. *Proceedings of the Royal Society of London—Series B: Biological Sciences*, 258, 93–9.

PIANKA, E. R. 2000. *Evolutionary Ecology* (6th edition). Menlo Park, CA: Benjamin Cummings.

PLATH, M., MAKOWICZ, A. M., SCHLUPP, I., & TOBLER, M. 2007. Sexual harassment in live-bearing fishes (Poeciliidae): comparing courting and noncourting species. *Behavioral Ecology*, 18, 680–8.

PTACEK, M. B. & TRAVIS, J. 1996. Inter-population variation in male mating behaviours in the sailfin mollie, *Poecilia latipinna*. *Animal Behaviour*, 52, 9–71.

PUSEY, A. E. & PACKER, C. 1994. Infanticide in lions: consequences and counterstrategies. *In*: PARMIGIANI, S. & VOM SAAL, F. S. (eds.) *Infanticide and Parental Care*. Chur, Switzerland: Harwood Academic Publishers, pp. 277–99.

ROSENTHAL, G. G. 2017. *Mate Choice*. Princeton, NJ: Princeton University Press.

RUNKLE, L. S., WELLS, K. D., ROBB, C. C., & LANCE, S. L. 1994. Individual, nightly, and seasonal variation in calling behavior of the gray tree frog, *Hyla versicolor*: implications for energy expenditure. *Behavioral Ecology*, 5, 318–25.

RYAN, M. J. 1985. *The Tungara Frog: A Study in Sexual Selection and Communication*. Chicago, IL: University of Chicago Press.

RYAN, M. J., HEWS, D. K., & WAGNER, W. E. 1990. Sexual selection on alleles that determine body size in the swordtail *Xiphophorus nigrensis*. *Behavioral Ecology and Sociobiology*, 26, 231–7.

SCHACHT, R., RAUCH, K. L., & MULDER, M. B. 2014. Too many men: the violence problem? *Trends in Ecology & Evolution*, 29, 214–22.

SCHARTL, M., ERBELDING-DENK, C., HOLTER, S., NANDA, I., SCHMID, M., SCHRODER, J. H., & EPPLEN, J. T. 1993. Reproductive failure of dominant males in the poeciliid fish *Limia perugiae* determined by DNA fingerprinting. *Proceedings of the National Academy of Sciences of the USA*, 90, 7064–8.

SCHLUPP, I., MCKNAB, R., & RYAN, M. J. 2001. Sexual harassment as a cost for molly females: bigger males cost less. *Behaviour*, 138, 277–86.

SCHÜRCH, R. & TABORSKY, B. 2005. The functional significance of buccal feeding in the mouthbrooding cichlid *Tropheus moorii*. *Behaviour*, 142, 265–81.

SHUSTER, S. M. & WADE, M. J. 2003. *Mating Systems and Strategies*. Princeton, NJ: Princeton University Press.

SIMMONS, L. 1992. Quantification of role reversal in relative parental investment in a bush cricket. *Nature*, 358, 61–3.

SIMMONS, L. & BAILEY, W. 1990. Resource influenced sex roles of zaprochiline tettigoniids (Orthoptera: Tettigoniidae). *Evolution*, 44, 1853–68.

SOMMER, S. 2000. Sex-specific predation on a monogamous rat, *Hypogeomys antimena* (Muridae: Nesomyinae). *Animal Behaviour*, 59, 1087–94.

STONE, E. A., SHACKELFORD, T. K., & BUSS, D. M. 2007. Sex ratio and mate preferences: a cross-cultural investigation. *European Journal of Social Psychology*, 37, 288–96.

SZÉKELY, T., WEISSING, F. J., & KOMDEUR, J. 2014. Adult sex ratio variation: implications for breeding system evolution. *Journal of Evolutionary Biology*, 27, 1500–12.

TOBLER, M., FRANSSEN, C. M., & PLATH, M. 2008. Male-biased predation of a cave fish by a giant water bug. *Naturwissenschaften*, 95, 775–9.

TREXLER, J. C., TEMPE, R. C., & TRAVIS, J. 1994. Size-selective predation of sailfin mollies by two species of heron. *Oikos*, 69, 250–8.

TREXLER, J. C., TRAVIS, J., & MCMANUS, M. 1992. Effects of habitat and body size on mortality rates of *Poecilia latipinna*. *Ecology*, 73, 2224–36.

VILLARREAL, A. E., GODIN, J. G. J., & BERTRAM, S. M. 2018. Influence of the operational sex ratio on mutual mate choice in the Jamaican field cricket (*Gryllus assimilis*): testing the predictions of the switch point theorem. *Ethology*, 124, 816–28.

WARNER, D. A. & SHINE, R. 2007. Reproducing lizards modify sex allocation in response to operational sex ratios. *Biology Letters*, 3, 47–50.

CHAPTER 7

Ornaments in Females

7.1 Brief outline of the chapter

In this chapter I want to explore the role of ornamental traits in females. They pose a bit of a conundrum, as they are not really predicted to exist—at
least until recently. From a simple sexual selection
point of view, female ornaments should be selected
against by males because inconspicuous females
suffer less predation and are more likely to care successfully for their offspring. Yet, countless species
show ornaments in females. Are they adaptations of
some kind or just the side effect of a genetic correlation? And what information do female ornaments
convey to males?

7.2 Introduction

By now I have hopefully established that males often
choose their mates, as do females. Consequently,
females might be expected to have ornaments that
evolve under male mate choice and compete for
males whenever this increases their fitness. However,
actually ornamentation in females is not that easy to
understand. Females are predicted to benefit from
being inconspicuous, and males should benefit
from preferring dull females. Yet, females of many
species have traits that must be considered ornamental. What can explain this? One general idea is
that female ornaments play a role in female–female
competition, which I will cover in more detail in
Chapter 8. In the current chapter I will instead
focus on female ornamentation. As I argued in
Chapter 1, male mate choice, female ornamentation, and female competition interact and influence
each other. Another consequence of the co-existence
of male and female choice is that these processes
happen at the same time and lead to some form of

mutual mate choice. This will also be addressed
separately (Chapter 9).

Males have traits that are considered ornaments, or
"conspicuous secondary sex traits" (Andersson,
1994), which are often or mainly under sexual selection, but can be counter-selected by natural selection. In other words, these traits are costly to males
and can best be understood if we assume the cost is
balanced by an advantage to males via sexual selection. This connects the concept of ornaments tightly
to the idea of traditional sex roles and it only works
if we assume that the bearer of ornaments is somewhat dispensable. To cite a classic example, male
peacocks risk being eaten by a fox, but those that
survive have more offspring. In a way, males are
dispensable. This is not the case for females, which
in some species take care of the young and are not
dispensable. Consequently, arguing that male mate
choice drives female ornaments is difficult.

A somewhat related phenomenon might be the
puzzling fact that males sometimes harm the females
they mate with (Chapman et al., 1995), which is
found for example in *Drosophila*. In this species,
males transfer components in their seminal fluid
that are costly to females. This has been interpreted
as negative pleiotropic effects linked to other traits
that are otherwise adaptive (Morrow et al., 2003),
but the bottom line seems to be that males could be
freer to evolve costly ornaments because they are
not crucially important to raising the offspring.
Females, on the other hand are in many species
indispensable to the offspring. This would predict
that females should be selected to survive, but not
to have ornaments. Yet, in principle the same mechanisms identified for the evolution of ornaments in
males may operate in females (Clutton-Brock, 2009).

Male Choice, Female Competition, and Female Ornaments in Sexual Selection. Ingo Schlupp, Oxford University Press (2021). © Ingo Schlupp (2021).
DOI: 10.1093/oso/9780198818946.003.0007

Clearly, there are ample examples of "conspicuous secondary sex traits" (Andersson, 1994) or "decorative traits" (Amundsen, 2000) in females. This is a somewhat serious conflict between a general prediction and reality. Intuitively, females should be better off investing the energy spent on an ornament in their offspring, and this is captured by the different Bateman gradients found in males and females (see Chapter 2) (Bateman, 1948). In principle the same argument would hold for males, but here we understand that males need to advertise their quality to females. Females, on the other hand, should be drab and inconspicuous, but often they are not. What if ornaments evolve in females because they too benefit from advertising to the choosing sex, in this case males? Consequently, if females advertise with their ornaments, the fitness gain of doing so must be greater than the cost of signaling (see Chapter 1). This is not easy to demonstrate and often we still know little beyond the existence of what looks like an ornament in females. Even studies of male responses to such ornaments are rather rare. And for such studies one could predict that males should reject showy females, as they might be less likely to survive to actually raise the offspring. But of course, the ornament observed may not only play a role in mate choice, but also double as an armament in female competition (Chapter 8).

Several hypotheses have to be considered regarding the evolution and adaptive value of ornaments in females and they are discussed in the sections below. First, they may not really be ornaments and have evolved under natural selection. Males and females may experience different selection regimes leading to differences between males and females that are not related to male mate choice. Second, there might be some genetic correlation between an ornament expressed in males and a version of that in females. In this case, the ornament may not be advantageous to the female, but females may still bear some of the cost of displaying the ornamental trait. Third, males may favor ornaments in females and cause sexual selection on females, akin to the sexual selection caused by female choice (Clutton-Brock, 2009), but probably weaker. Finally, social selection (West-Eberhard, 1979, 1983; West-Eberhard, 1989) provides a framework for understanding trait evolution in females (Lyon and Montgomerie, 2012; Tobias et al., 2012).

7.3 Potential origins of female ornaments

Although their function and origin may not be entirely clear, there can be no argument that females of many species have traits that look like ornaments or showy traits. Many bird species for example, show spectacular color ornaments. Consider the bright yellow crest in males of the sulphur-crested cockatoo (*Cacatua galerita*). The crest stands out in this all-white bird. It seems reasonable to assert that the crest is an ornament and might have evolved under sexual selection. This assertion, however, is made more complicated by the fact that females have a similar crest. Indeed, in many bird species both sexes are very colorful, but monomorphic. Much to the frustration of ornithologists and birders, for example most seabirds are monomorphic, yet often quite colorful. Take the puffin, *Fratercula arctica*, a very colorful bird, with an especially brightly colored beak. If one would see only one exemplar, one would be tempted to ascribe the colorful beak to sexual selection, but the sexes are so similar that a statistical technique, discriminatory function analysis, is needed to help sex the birds (Bond et al., 2016). There are differences—mainly in size—between males and females, but they are very subtle (Doutrelant et al., 2013). Puffin beaks also have a fluorescent color component to them (Dunning et al., 2018) that humans completely miss, because we are unable to see in the ultraviolet (UV) spectrum. But this UV fluorescence is also monomorphic. In general, as more species are investigated, subtle differences between the sexes may be discovered in more species (Morales et al., 2020), but overall, a good number of bird species are monomorphic. Similar patterns are found in other groups, as unrelated as penguins (Figure 7.1) and shorebirds: the sexes are monomorphic, but colorful. One example is the gentoo penguin, *Pygoscelis papua*, which has a red beak and a white stripe on the head.

In another species of penguin, the yellow-eyed penguin (*Megadyptes antipodes*), which is also monomorphic, the color of the eye was positively correlated with parental ability (Massaro et al., 2003). One

Figure 7.1 Gentoo penguin (*Pygoscelis papua*) (photo credit: Ingo Schlupp).

Figure 7.2 Jamaican tody (*Todus todus*).
Source: Wikimedia Commons (https://commons.wikimedia.org/wiki/File:Todus_todus_cropped.jpg; photo credit: Dominic Sheroni. Reproduced under Creative Commons Attribution-Share Alike 2.0 Generic license).

further example is the Jamaican tody (*Todus todus*), where both sexes sport a bright red chin and beak (Figure 7.2). They are easily spotted in their typical

forest habitat. Unless this actually *is* the function of the red chin, it is difficult to imagine what the benefit of being so conspicuous is for a small bird.

Indeed, in birds the phylogenetic mix of patterns is amazing. In many ducks, males are highly ornamented while females are relatively dull and inconspicuous, and presumably less detectable by their predators. They may, however, have some traits that resemble ornamental traits in males, like a few blue feathers in female mallards (*Anas platyrhynchos*) (Figure 7.3). In the closely related geese and swans, most species are quite monomorphic, but can be surprisingly colorful like the Egyptian goose (*Alopochen aegyptiaca*) or have conspicuous markings like the bar-headed goose (*Anser indicus*). Again, this is not to say that there is absolutely no sexual dimorphism, as in bar-headed geese, for example, males are slightly larger (Lamprecht, 1986), although similar in coloration. Overall, however, if both sexes appear to be relatively cryptic, most likely this is because of natural selection, and easily explained. Similarly, in cases where males are colorful and females not, we invoke sexual selection, often via female choice. If males lack ornamentation, but females are colorful, this may be associated with sex-role reversal such as in *Phalaropes*. The cases that require explanation are the ones where both sexes are ornamented and especially cases where females show ornamental traits unique to them. As a mechanism for dichromatism loss of pigmentation owing to small genetic differences has been suggested (Gazda et al., 2020). So far, we have only considered bird examples, but similar patterns are found in many other taxa.

Before we turn to a more detailed analysis, however, we need to consider a few corollaries. First, the examples presented so far are based on human perception. And most examples include only visual signals, ignoring all other sensory systems. By doing this we run the risk of arriving at wrong conclusions. Even considering only visual communication, we know that most species have visual systems that are different from human vision. In birds a key difference is that most species are tetrachromatic with four color cones (Vorobyev et al., 1998) allowing them to perceive the large number of colorful bird plumages (Stoddard and Prum, 2011), whereas humans have three cones. One of the color cones in

Figure 7.3 Mallard (*Anas platyrhynchos*) (photo credit: Ingo Schlupp).

birds is sensitive in the UV spectrum, a part of the spectrum that humans cannot perceive. In other words, it is very possible that some of the species we consider to be monomorphic have differences that are just not visible to humans. Another example has been reported in a jumping spider, where only males show UV reflectance (Lim and Li, 2006). Furthermore, color needs to be considered in an ecological and sensory context. A yellow reef fish may appear very colorful to humans when seen in a glass aquarium, but may be perfectly camouflaged in its natural environment, depending on the visual conditions there. The environment of our reef fish might be dominated by yellow corals (Marshall, 2000), making bright yellow perfectly cryptic. There is an interesting story associated with the discovery of UV vision in birds. In a review paper Bennett remarked in 1994 that "almost without exception, such studies have ignored the major differences between human and avian color vision" (Bennett et al., 1994). Ultraviolet light occurs in the spectrum between frequencies of 10 and 400 nm and many organisms respond to UV light. Humans, for example, cannot see in the UV but respond to UV with an increase of melanin in skin cells when stimulated by UV. Many other animals can see in the UV. Perhaps best known are the classic experiments by Nobel Laureate Karl von Frisch on bees (Munz, 2016). But well before von Frisch, actually in 1882, UV vision in insects had been reported (Lubbock, 1882; Clark, 1997). That some vertebrates have UV vision has also been known for a long time, based on experiments by Schiemenz on minnows (*Phoxinus phoxinus*) (Schiemenz, 1924). This made it likely that UV vision was present in other vertebrates as well. Yet, it took a long time until UV vision in birds was discovered by two groups independently in May (Wright, 1972) and October of 1972 (Huth and Burkhardt, 1972). One study, by Wright, used pigeons (Wright, 1972); the other, by Huth and Burkhardt, a hummingbird (Huth and Burkhardt, 1972). UV vision was detected using elegant behavioral experiments; in Wright's case much to the surprise of the author. However, these important findings were unknown to or ignored by scientists in behavioral and sensory ecology and many behavioral experiments were conducted on color-based mate choice without including UV vision in the experimental design (Bennett et al., 1994). Eventually, this was criticized as problematic and anthropomorphic in a seminal review in 1994 (Bennett et al., 1994) and since then studies have taken a more animal-focused approach (Tanaka, 2015). This story leads to several conclusions: first it is important to pay attention to what colleagues in neighboring disciplines publish, hard as it may be. A second conclusion is that findings will be ignored if not published in the *lingua franca* of the time. The paper by Huth and Burkhardt was published in a prominent journal, but in German. At that time English had already replaced German as the *lingua franca* of science.

In essence, color—and any other potentially ornamental trait—can be under selection for many different reasons and there is no *a priori* reason to assume that such selection has to work uniformly on both sexes. Indeed, dichromatism can evolve either by males losing color or females gaining color (Price and Birch, 1996). In other words, natural selection and not only sexual selection can lead to differences between the sexes (Badyaev and Hill, 2003). An especially good example is provided by another parrot, *Eclectus roratus*, which has a very distinct sex difference in coloration, but in this case—at least to the human eye—the males appear green, and more cryptic, whereas the females are bright red and blue and very conspicuous. This kind of reversed dichromatism is often associated with sex-role reversal, but this is not the case here (Heinsohn et al., 2005). Instead it appears that males and females are under completely different selection pressures leading to their coloration (Heinsohn et al., 2005). This is also found in other parrots and this guides us to the more general conclusion that ecology is driving this mating system. This means that the underlying genetic architecture for coloration is not coupled and evolves independently. There seems to be no genetic correlation. Modeling how male and female are seeing each other and how they are seen by potential predators based on

visual contrast in various situations, Heinsohn et al. (2005) found that males are conspicuous to other males when it matters, but cryptic to raptors. Male color appears to be a compromise of multiple functions. It seems as if the scarcity of nest cavities led to a polygyandrous mating system with female competition for nesting opportunities (Heinsohn, 2008). Females appear to have evolved to be more colorful because they spend most of their time inside a nest cavity, and their color may be important in intra- and interspecific competition for nesting cavities. But a role for male or female choice appears to be unknown (LeBas, 2006). Such female competition is probably widespread—not just for nesting opportunities—and will be discussed in more detail in Chapter 8.

Similarly, physical differences between habitats can select for differences in ornaments. In a fish from the Amazon region, the sailfin tetra (*Crenuchus spilurus*), differences in the light conditions in different types of water predict the degree of ornamentation (Pires et al., 2019). Taken together, this shows that sexual dimorphism is often, but not always driven by sexual selection. Conversely, a supposedly classic example for crypsis, the ability to change color in chameleons, may have a component of sexual selection. In a group of chameleons, the African dwarf chameleons (*Bradypodion*) signaling in a sexual selection context has been found to drive the evolution of color change, but not crypsis (Stuart-Fox and Moussalli, 2008).

With regard to color, many other functions and resulting selection pressures are known, including aposematism, mimicry, and crypsis, as well as color being an indicator of parasites and health in general (Hamilton and Zuk, 1982). A similar multitude of adaptive functions can be assumed for other traits, as well. Color is hardly unique in this way. All of these functions, of course, can drive selection simultaneously. Just one example is the role of dichromatism relative to nest predation (Martin and Badyaev, 1996). Another example is the finding that bird coloration seems to follow large-scale biogeographical rules, which predict that birds are darker in warmer and wetter areas (Gloger's rule) but also in colder areas (Bogert's rule) (Delhey et al., 2019).

In a few cases females have evolved to resemble male phenotypes. In a damselfly from Europe,

Ischnura elegans, for example, some females resemble males, likely to reduce sexual harassment (Cordero et al., 1998; Svensson and Abbott, 2005; Gosden and Svensson, 2009). The opposite case, males benefiting from resembling females, is much more common, usually as a strategy to avoid male competition and obtain sneaky matings (Oliveira et al., 2008). In essence this leaves us with multiple pathways for interpreting male and female ornamentation (Figure 7.4).

Female song is another example of ornamental traits in females—again looking mainly at birds. Somewhat overlooked for a long time (Riebel, 2003; Odom et al., 2014), it is increasingly clear that females of many bird species sing. Although I focus on birds here, I assume that female song is also found in other taxa that communicate acoustically, such as for example whales, frogs, and insects. There are some intriguing examples from female fishes that also sing (Bussmann et al., 2020). The sheer existence of female song makes it unlikely that all male song has evolved solely under sexual selection. Generally, more and more studies find that females of many birds actually do sing. In fact, the long ongoing debate about the functional role of duetting has highlighted the role of female song in many species (Langmore, 1998; Langmore and

Bennett, 1999), and suggests a role for social selection. As Price pointed out, sexual dimorphism is just a measure of difference between sexes and one has to be careful about making assumptions how this difference evolved (Price, 2015) (Figure 7.4). As illustrated above for instance with the parrot example, many possible scenarios for this dimorphism do not involve sexual selection, and natural selection and social selection may play a role. Price (2015) also remarks that our understanding of male and female song may reflect yet another bias, this time for studying temperate passerine birds, many of which are indeed dimorphic for song (Price, 2015). Another example might be female song in blackbirds (Whittingham et al., 1992, , 1996). In at least one species, the Australian magpie (*Gymnorhina tibicen dorsalis*), females actually sing more than males (Dutour and Ridley, 2020). In a wider context, the fact that we overlooked female song and its potential functions for a long time might be another example of gendered science (Ahnesjö et al. 2020). An overview of female bird ornaments can be found in Doutrelant et al. (2020).

With these cautionary notes in mind I want to discuss a few scenarios that could explain the evolution of female ornaments.

7.4 Males exercise sexual selection and females evolve ornaments

If males are choosy, they may induce the evolution of ornaments in females (Clutton-Brock, 2009). This would in essence mirror the process known to select for ornaments in males via female choice. Consequently, females should advertise their quality to the choosing males. But this might create a tradeoff for the females. The investment in their ornament is lost as an investment in offspring or other traits, raising the question of how ornaments evolve (see Chapter 1). In principle, of course, a similar argument could be made for males, but in many species their contribution beyond sperm is small. Consequently, females are predicted not to have ornaments, or very weak ones. Given this theoretical prediction, it is somewhat surprising that females of many species have quite conspicuous traits that could be ornaments and presumably come with some cost. In this context, ornaments

sexes monomorphic both cryptic Natural selection	**sexes dimorphic** males showy/females cryptic Sexual and natural selection
sexes monomorphic males and females showy Social selection	**sexes dimorphic** males cryptic/females showy Social selection

Figure 7.4 Possible explanations for the origin of female ornaments.

that are found in females only, and not in males, are of particular interest. Testing this hypothesis is essentially done in the same way we test the evolution of male ornaments: preference tests can establish which traits are under selection and ideally the fitness consequences of choice are also investigated. In a review of 43 studies in vertebrates looking at the relationship of offspring quality and ornamentation, Nordeide and colleagues (2013) found a somewhat mixed pattern: 20 studies reported a positive relationship (potentially rendering the ornament an honest signal), 9 showed a negative relationship, and 14 found no relationship. The sample was strongly biased toward studies using birds (Nordeide et al., 2013), but the findings are telling nonetheless: there is no clear pattern and we may be missing something important about the evolution of ornaments in females. Maybe mothers with stronger ornaments have offspring—male and female—that are more attractive to their mates. This would be an extended version of the "sexy son" hypothesis (Prokop et al., 2012; et al., 2015), and could be dubbed the "sexy daughter" hypothesis.

One interesting trait that is found only in females is the genital swelling in some primates, such as chimpanzees and several baboons. This is not only a fascinating example of a female trait that males may use to obtain information about a potential mate, it also highlights the complexity of the co-evolution of trait and preference. Motherhood is certain in all livebearing species. The individual that gives birth to the offspring is the mother. This is such a strong relationship that it is even codified in law, for example in paragraph §1591 BGB of German law. This is not an unimportant distinction as in egg-laying species a female can be duped into investing into offspring that are not her own. This is exemplified in the many studies of inter- and intraspecific brood-parasitism, such as egg-dumping in water pipits (Reyer et al., 1997), geese (Weigmann and Lamprecht, 1991), some insects (Loeb, 2003), and many fishes with parental care (Taborsky, 1994). But in all livebearing species motherhood was always perfectly clear until the arrival of reproductive medicine and surrogate mothers in humans, and only in humans. Fatherhood, on the other side, is not, as captured in the old Latin proverb *"Pater semper incertus est"* (fatherhood is always uncertain).

Consequently, in species with internal fertilization and more than a single offspring, mixed broods are possible and often occur. In many primates, where typically a single offspring is born and they exist in multi-male/multi-female mating systems, females advertise their ovulation via well-developed genital swellings. According to one of several evolutionary hypotheses (reviewed by Zinner et al., 2004), these swellings have apparently evolved into an imprecise signal that provides males with some, but not perfect information (the graded signal hypothesis) (Nunn, 1999). The expression of the swellings is under hormonal control (Heistermann et al., 1996; Dixson, 2015) and interestingly olfactory signals do not seem to play a role in detecting ovulation (Zinner et al., 2004). This leads to uncertainty among males over paternity, which may serve to impede male harm to offspring that they did not sire. If a female copulates with multiple males around the time of ovulation, none of the males knows exactly who has sired the offspring if the signal is vague and ambiguous. If the ovulation advertisement was quite precise, the male that mated with the female right before ovulation could be certain to be the father. Consequently, if there is a chance (a probability larger than zero, but much smaller than one) that any given male of a group could be the father, the offspring is better protected against male aggression or infanticide. At the same time the male who is most likely to be the father is also likely to invest in the offspring. Sperm longevity in the female genital tract will also contribute to this uncertainty. This is different for example in lions (*Panthera leo*), where males taking over a pride of females can be certain that none of the offspring present at the time of takeover is theirs. Males then usually kill the young offspring, thereby inducing estrus in the females of that group (Pusey and Packer, 1994). In humans, ovulation is concealed and evidence for the ability of men to recognize ovulation in women is circumstantial (Buss, 2015). In other primates, such as baboons, there is some evidence that the size of swellings correlates with female quality (Domb and Pagel, 2001), but this argument is disputed by others (Zinner et al., 2002).

Phylogenies show that a majority of anthropoid primates have female genital swellings (Sillén-Tullberg and Moller, 1993). They have evolved and been lost several times and it appears that the absence of swellings is associated with the absence of male–male competition. There is also considerable variability in the trait and how precisely it indicates ovulation between closely related species like bonobos (Heistermann et al., 1996; Douglas et al., 2016) and chimpanzees (Deschner et al., 2003, 2004). Looking at the bigger picture, this shows again how relevant ecology and mating systems are in the evolution of male mate choice and female ornaments. It is also important to note that there is a lively debate around the function of genital swelling among primatologists (Zinner et al., 2004). The main message for this book, however, is that genital swellings are a signal and ornamental trait that has evolved in females and to which males generally respond in a meaningful way (Zinner et al., 2004; Emery Thompson and Wrangham, 2008). The critical question is whether male mate choice has driven the evolution of the swellings. This is absolutely possible, but far from clear.

In modern humans it is fairly obvious that women show traits that are attractive to men. These traits are thought of as ornamental and may relay information that men use to estimate fecundity, such as the waist to hip ratio (Buss, 2015) and related measures (Tovée et al., 1999; Fan et al., 2004). The skin tone and the color of the lips have also been suggested as ornaments (Thornhill and Grammer, 1999), but whether these traits really are ornaments is not clear. Lip color, for example, can be enhanced or altered artificially by using lipstick, and—maybe most profoundly—women and men can change their physique using plastic surgery or tattoos either permanently or by temporarily adding completely novel and artificial traits to their phenotype. Cosmetics are not only used to attract men, but also to compete with other women (Mafra et al., 2020). The cosmetics and fashion industries are based on this.

Artificial ornaments are very popular in humans, but not unique to us. They are also not restricted to ornaments using vision, but also occur in the form of perfumes, using chemical senses. In fact, quite a few species have evolved the use of artificial ornaments, perhaps displaying an extended phenotype (Dawkins, 1982). This is the case in bowerbirds, where males not only use objects from nature but also manmade objects to attract females to mate

with them (Patricelli et al., 2003). In at least one species of bowerbird the bowers are constructed in such a way that they use an optical illusion, the Ebbinghaus illusion, to attract females (Kelley and Endler, 2012). Interestingly, although some male bowerbirds are strongly ornamented, in other species, males are relatively drab, and their ornament seems to be mainly or solely their bower. This may reduce the risks associated with bodily showy ornaments, while retaining an ornamental trait to attract females, the bower. An example for this is McGregor's bowerbird (*Amblyornis macgregoriae*), where males have a colorful crest that is only visible while courting. I am not aware of examples of females producing extended phenotype ornaments like males. However, females are often building structures such as nests that might qualify.

Somewhat related, several species have been shown to respond to novel ornaments that normally are not found in that species and that have been artificially added by human experimenters. One example are zebra finches (*Taeniopygia guttata*), in which females sometimes prefer males with certain leg bands used for identifying individual birds (Burley et al., 1982). Later studies used this paradigm to study the role of artificial ornaments in social influences on mate choice (Kniel et al., 2015a, 2015b, 2016, 2017).

As a tangent, in fishes a number of studies investigated female, but not male, preferences for artificial traits that are not completely outlandish, but might have evolved or were present in an ancestral species. The context for this is the pre-existing bias concept (Ryan, 1990; Ryan and Keddy-Hector, 1992; Ryan and Rand, 1998; Endler and Basolo, 1998). Some of this work was done in livebearing fishes, but other examples exist. Starting in 1990, Basolo investigated how females respond to male traits found in some species, but not in others. Concretely she investigated the evolutionary sequence of how the sword in some members of the genus *Xiphophorus* evolved, testing the hypothesis that the evolution of female preferences for the trait is predating the evolution of the trait (Basolo, 1990, 1995, 2002; Makowicz et al., 2016). In two species she also looked at male responses to the male ornament in females. Male *Xiphophorus hellerii*, a species that naturally shows a sword in males, did not prefer females of their own species artificially displaying a sword. By contrast, males of *Priapella olmecae*, representing a genus that is the sister taxon to *Xiphophorus*, did prefer females with an artificial sword. This indicates that males may have a pre-existing preference for ornamental traits in females (Basolo and Delaney, 2001). Another study explored this trait and preferences for it in more distantly related species, the mollies of the genus *Poecilia*, and found preferences in some, but not all (Makowicz et al., 2016). Most interestingly, they studied the Amazon molly (*Poecilia formosa*), which is a sperm-dependent unisexual hybrid (Schlupp, 2005). The question asked was whether the Amazon mollies would show the ancestral preference of the paternal or the maternal ancestor. Instead they found that Amazon mollies show the preferences of the host species they live with (Makowicz et al., 2016). Looking at another trait, a mustache-like growth found in some mollies (Schlupp et al., 2010), using a phylogenetic approach (McCoy et al., 2011), we found that some species show a preference for the novel trait, but more importantly, in many species some individuals prefer the novel trait, which is a sufficient starting point of a novel trait to spread through a population. While all of the existing research is on female responses to novel and artificial ornaments, I think some work on male responses to artificial and novel female ornaments would be very useful. Also, more work on potential traits that may meet the definition of extended phenotypes (Dawkins, 1982) would be very important.

7.5 Ornaments in females and what they might mean

There are some taxon-specific trends that I want to address below. We have many examples for ornamental traits in birds. Birds are a great group to address questions about female ornaments because we know so much about ornaments in males and about birds in general.

A prominent example of male preference for female ornaments is found in bluethroats, *Luscinia s. svecica*, a northern European bird with distinct blue and red ornaments in both sexes. Males preferred females that were more colorful. In turn coloration in females was correlated with female body size, which may be indicative of female fitness (Amundsen et al., 1997). Another study, however, did not find a correlation of female plumage and maternal quality (Rohde et al., 1999). Another ornament only found in females is known from a marine duck, the common eider (*Somateria mollissima*). In this case it is a white stripe on their cover wind feathers and the expression of the stripe is correlated with indicators of female quality like mass loss. Clutch size and whiteness were also correlated (Hanssen et al., 2009).

In tanagers (Thraupidae), female ornamentation was studied in a phylogenetic context (Burns, 1998). While sexual selection was clearly a possible pathway to female ornaments, this was by no means the only one.

Interestingly, in a hummingbird, the amethyst-throated sunangel (*Heliangelus amethysticollis*) (Bleiweiss, 1992, 1997), female coloration seems to have evolved relative to female competition for nectar. Female coloration in other hummingbird species seems to be because of factors other than sexual selection (Bleiweiss, 1992). Males and females are colorful in many species but occupy different phenotype spaces.

In yet another bird, the red-winged blackbird (*Agelaius phoeniceus*) from North America, males and females have colored epaulets. The male red ornament is expressed much stronger and very bright, but in females brightness is correlated with condition, and possibly plays a role in female competition for nest sites (Johnsen et al., 1996). However, a study using dummies in red-winged blackbirds found no correlation between epaulette coloration and any fitness relevant trait (Muma and Weatherhead, 1989).

In barn swallows (*Hirundo rustica*) males and females have elongated tail feathers. Manipulating these streamer feathers in females leads to assortative mating (Cuervo et al., 1996), but in unmanipulated females the length of the streamers was not an indicator of fecundity or ability to provide for the offspring. Furthermore, an earlier study had already shown that female ornaments in barn swallows are correlated with female reproductive potential (Moller, 1993).

Another example comes from house finches (*Carpodacus mexicanus*), where females can show a less intense version of male coloration. Males prefer females with brighter coloration, but female age is more important (Hill, 1993). Because there was no correlation with several traits that might be indicative of individual female quality, such as overwinter survival, reproductive success, or condition, Hill concluded that the male preference might be a correlated response, and reflect the well-known preference in females for colorful males in house finches (Hill, 2015). In the rock sparrow, *Petronia petronia*, males and females both have a colorful patch on the breast. Males prefer females with bigger ornaments, without a clear benefit (Griggio et al., 2005, 2009), which may be explained by indirect benefits.

In zebra finches, *Taeniopygia guttata*, both males and females show red beaks, but the color is much more intense in males. Another study found that males and females both show preferences for beak color, but while female preferences were directional and they preferred the brightest beaks, male preferences were stabilizing and favored mid-range colored beaks (Burley and Coopersmith, 1987). In this species redder bills are advantageous for males, but for females, redness comes with a cost. For example, they experience earlier mortality (Price and Burley, 1994). This is an interesting case of opposing selection on what might be a correlated ornament. Males choose females that are more fecund (Monaghan et al., 1996). In the red junglefowl, *Gallus gallus*, males prefer females with bigger ornaments, which are correlated with higher fecundity (Cornwallis and Birkhead, 2007). Males also invested more sperm in females that had bigger ornaments.

Crested auklets have been found to show mutual sexual selection (Jones and Hunter, 1993; Jones and Hunter, 1999). In northern cardinals (*Cardinalis cardinalis*), a finch from North America, where females show a spectacular red plumage—which again is not as bright as male plumage but still quite conspicuous—female plumage was found to be an indicator of female feeding activity (Linville et al., 1998). Yet another example comes from a bird of prey: in lesser kestrels, *Falco naumanni*, a small raptor from southern Europe and Asia, females can show male plumage traits such as gray rumps. Male coloration in females is not correlated with female fitness and because it increases with age, changes in

hormones were suggested as the proximate cause (Tella et al., 1997). Finally, in dotterels (*Charadrius morinellus*) females are more intensively ornamented than males, and although the species appears to be sex-role reversed, there is no clear indication of male mate choice (Owens et al., 1994).

Overall, it appears as if colorful plumages in female birds have evolved under a number of scenarios, and sexual selection is just one of them. One obvious, but somewhat trivial, conclusion from the available studies is that not every dimorphism found is always due to sexual selection on males. More detailed studies are needed to understand the adaptive value of female ornaments in birds (Doutrelant et al., 2020).

Another group where some work has been done on female ornaments is reptiles, especially lizards. In this group we also know a little more about mechanisms, partly because of the great tradition of studying hormones and behavior in this group (Hews et al., 1994; Adkins-Regan, 2005). For example, female common chameleons (*Chamaeleo chamaeleon*), change color patterns in synchrony with their reproductive status throughout the breeding season, rendering their color display into a trait that males may use as indicator (Cuadrado, 2000). Males in this species preferentially guard larger females (Cuadrado, 1998).

A study investigating an agamid lizard (*Ctenophorus ornatus*) found that female throat color was not correlated with female quality *per se*, but with female laying date and hence may indicate female receptivity. Males strongly preferred females with higher throat color chroma (LeBas and Marshall, 2000), although the indicator value of it is unclear.

In fence lizards, *Sceloporus undulatus*, females can show reduced versions of male ornaments. This ornament is a male-like ventral color pattern that gave them the nickname "bearded ladies." Males rejected these ornamented females (Swierk and Langkilde, 2013; Swierk et al., 2013). Furthermore, the ornamented females were found to have lower reproductive output, and their eggs were laid later and hatched later (Swierk and Langkilde, 2013). I believe these findings are in agreement with the idea of a cost to having an ornament owing to a genetic correlation. But there may be compensatory advantages for ornamented females, as they were found to show higher sprint speeds, which may be an advantage in predator escape behavior (Assis et al., 2018).

In tropidurid lizards (*Microlophus occipitalis*), females can have red throats, which are reminiscent of male ornaments (Watkins, 1997). In this case throat color correlated with reproductive state and females that were about to lay eggs or had recently laid a clutch showed more coloration. Watkins suggested that females might signal to males that they are not interested in mating. This was supported by behavioral data showing that males preferred less ornamented females (Watkins, 1997).

By contrast, there is a correlation between offspring quality and ornamentation in another fence lizard, *Sceloporus virgatus* (Weiss et al., 2009; Weiss and Dubin, 2018). Males respond to female ornaments in various ways, all in agreement with the notion that ornamented females are preferred. For example, males compete stronger for ornamented females (Weiss and Dubin, 2018).

There are a few amphibian examples of female ornamentation. Just as in birds, amphibian females can be very colorful. Sometimes this is easily explained, for example in Dendrobatids, where natural selection confers an advantage of displaying warning coloration to both sexes. A similar argument has been proposed for the yellow and black pattern found in European fire salamanders (*Salamandra salamandra*), which are also toxic and distasteful, and some toads (*Bombina bombina*). In many species of the genus *Ichthyosaura* (formerly *Triturus*), a group of European newts, both males and females have colorful undersides, which are normally not visible and may be flashed to potential predators in what is called the "Unkenreflex." In the Alpine newt, *Ichthyosaura alpestris* both sexes have a bright red to orange belly and females with more orange are preferred by males and have higher fecundity (Lüdtke and Foerster, 2018, 2019) (Figure 7.5). In a study on wood frogs (*Lithobates sylvaticus*), males showed a preference for females in a binary choice test, but when mated to their nonpreferred females those offspring had higher fitness (Swierk and Langkilde, 2019). In an interesting parallel to song in female birds, there are a few anurans with female calls. We already encountered this in Chapter 3, when we talked about midwife toads,

Figure 7.5 Alpine newt female (right) (*Ichthyosaura alpestris*). Male on the left (photo credit: Deike Lüdtke).

Alytes, but there are more examples (Emerson and Boyd, 1999). In the few cases reported, it seems that female calls are produced under the influence of male hormones and around the time when females have to lay their eggs. Instead of losing their investment, they might be trying to actively attract males. One wonders how many examples of female vocalizations have been overlooked. How easily this happens can be seen in this example: common wisdom held that males of the common toad (*Bufo bufo*) do not call, and simply wait for females at the margins of breeding ponds to first intercept and then breed with them. However, males do call, but only at low density (Höglund, 1988), which is easily missed by field researchers (see also Chapter 6).

There are some additional vertebrate examples coming from fishes. As predicted ornaments are found in females of sex-role reversed species (Bernet et al., 1998), but that is not the focus here. More relevant is that also males of sex-role reversed species may have ornaments (Matsumoto et al., 2010). A particularly interesting ornament was found in males of a pipefish species from Japan, *Corythoichthys haematopterus*. This species is sex-role reversed (Matsumoto and Yanagisawa, 2001; Sogabe and Yanagisawa, 2007), so an ornament in a male is viewed as the equivalent of an ornament in a female in a non-sex-role reversed species. In this particular case, the males show a number of blue and yellow spots in the head and thorax region. The spots seem to be condition dependent (Matsumoto et al., 2010), but how females respond to them is not known. The fishes form lifelong pairs with no extra-pair copulations, hence there may be little opportunity for mate choice (A. Sogabe, pers. comm., June 20, 2019), although mutual mate choice seems possible.

A comparative study of two closely related families of fishes revealed that sexual selection is involved in the evolution of dichromatism in one group, but not the other (Méndez-Janovitz et al., 2019). The authors used two families of fishes, Poeciliinae and Goodeinae, both of which show ornamental traits in males and females. They found that dichromatism, the difference in coloration between the sexes, is likely driven by female choice in Poeciliinae, but suggest that male choice played a role in the evolution of female coloration in Goodeinae. A previous study had already reported male mate choice for female size and coloration in a goodeid fish, *Girardinichthys viviparus* (Méndez-Janovitz and Macías Garcia, 2017).

Sockeye salmon (*Oncorhynchus nerka*) males and females display red coloration, and males show a preference for redder females (Foote et al., 2004). It is not clear if female redness is correlated with traits that indicate female quality. However, the idea that the expression in females is based on a genetic correlation with expression in males has been challenged by a study in Arctic char (*Salvelinus alpinus*), mainly because expression of carotenoid-based coloration as a mere side effect would be so costly to females that it would be likely to become uncoupled in evolution (Skarstein and Folstad, 1996).

In brook sticklebacks, *Culaea inconstans*, females have multiple ornaments (McLennan, 1994), and males seem to base their preferences for females on those (McLennan, 1995). In the European stickleback, *Gasterosteus aculeatus*, females display red bellies, which is used in male mate choice. Because the carotenoids used in this ornament have to be acquired through food, they are limited, and females have to tradeoff investing carotenoids in either eggs or in their ornament at the base of the spines of the pelvic fins (Nordeide et al., 2006). This is interesting and somewhat counterintuitive because males that choose more colorful females fertilize eggs that have fewer beneficial carotenoids (Nordeide et al., 2006).

In *Gobiusculus flavescens*, a small marine fish found in Sweden, Amundsen and Forsgren studied male preferences. This species is not sex-role reversed, but females show a distinct color ornament, not found in males, a bright orange belly. Males strongly prefer females that display more colorful bellies, because these bellies are possibly correlated with female fecundity (Amundsen and Forsgren, 2001, 2003). In a goby species from Japan, *Rhinogobius* sp., nuptial coloration as an ornament in females was described, but its role is not clear (Takahashi, 2000).

The final group I want to look at is insects. In a dance fly, *Rhamphomyia tarsata*, females honestly advertise fecundity to males with pinnate scales (LeBas et al., 2004). In this mating system males provide nuptial gifts, and females are usually ornamented, whereas males are drab. In this situation females might actually benefit from deceiving males about their fecundity and thus obtain more nuptial gifts, but this is not what they do. Although this species is not sex-role reversed, females do compete for males and their nuptial gifts, and their ornaments may play a role in female competition. In a relative, an undescribed fly of the same genus, *Rhamphomya*, females appear to show lekking behavior, have ornamented legs, and males seem to choose females (which makes this species sex-role reversed), yet interestingly males also provide nuptial gifts, the equivalent of an ornament (Alcock, 2016). Female competition for nuptial gifts in this group is widespread (Hunter and Bussière, 2019). In a study using wild stalk-eyed flies (*Teleopsis dalmanni*) eyespan was found to be a good indicator for female and male reproductive quality (Cotton et al., 2010).

A study of male mate choice for a potential female ornament in an Australian butterfly (*Aeschynomene indica*) found only weak preferences. In this case the female trait is a wing patch that reflects in the UV (Rutowski and Kemp, 2017). Male choice has also been described in Australian field crickets (*Teleogryllus oceanicus*). Here males can base their preference on a cocktail of cuticular hydrocarbons (CHC), and some components correlated with female fecundity (Berson and Simmons, 2019).

7.6 Genetic correlation

Another reason why females might be ornamented is some form of genetic correlation. Clearly, the idea that ornaments show up in females because they evolved in males is very intuitive and quite possible (Lande, 1980). For example, in a mate choice study using dummies to find out if males and females preferred the colorful epaulettes in red-winged blackbirds (*Agelaius phoeniceus*), a genetic correlation was proposed as the most likely basis for the existence of the ornament (Muma and Weatherhead, 1989). The knowledge on the genetic architecture underlying expression of ornaments in females has been summarized by Kraaijeveld (Kraaijeveld, 2014). A large-scale study of bird color ornamentation found that selection on both male and female plumage can operate additively and also differentially on both sexes (Dale et al., 2015). Although a genetic correlation cannot be ruled out, it can clearly be overcome and broken up by selection. Several patterns detected in this context are worth further consideration, for example that cooperative breeding and living in the tropics may select for female coloration, likely tied to strong resource competition (Dale et al., 2015).

There is very strong evidence for the expression of male genes in females from hormone treatment studies. It seems relatively easy to reach a male phenotype from a female phenotype. Feminization, the expression of female genes in males, is also possible and sometimes associated with hormone mimics as environmental pollution (Hayes et al., 2002), as reported for several amphibians. In some species the hormonal basis of the expression of male-like coloration and behavior in females has been investigated. For example, in the eastern fence lizard, *Sceloporus undulatus*, treatment with testosterone was able to induce male-like coloration in juvenile females (Cox et al., 2005).

An extreme example comes from a livebearing fish, the all-female Amazon molly (*Poecilia formosa*) (Schlupp, 2005), where male genes were thought to be obsolete, because only females are known from this species. However, treatment with male hormones, in this case methyltestosterone, revealed that in this all-female species male traits could be expressed. Hormone-treated females showed male-like color patterns, growth of a male-specific modified fin, the gonopodium, which is used for sperm transfer, and spermatogenesis (Schartl et al., 1991; Schlupp et al., 1998). Such masculinized females

also showed male-like behavior (Schlupp et al., 1992). Other studies have begun to look into the genomic basis of the retention of apparently obsolete genes in this fish (Zhu et al., 2017; Schedina et al., 2018; Warren et al., 2018). This lends indirect support to the possible existence of genetic correlations between male and female ornaments. Yet, as argued by Dale and colleagues, such a correlation should be easily overcome under selection (Dale et al., 2015).

Finally—after criticizing the concept of a genetic correlation as explanation for female ornaments—one could use the same logic to explain female preferences for larger males in cases where there is no clear benefit to females (Ryan and Keddy-Hector, 1992). This could be because of a genetic correlation with the well-documented strong preferences found in males for larger females. This preference is relatively easy to understand and is supported by the widely assumed fecundity benefit (Schlupp, 2018) (see Chapter 3). Males choosing larger females will have more offspring because larger females are more fecund. To my knowledge the idea of a genetic correlation as explanation for the widespread female preference for larger males still awaits testing. To be clear, in many cases there is a clear benefit to mating with larger males, but in other cases there is no clear benefit and good genes arguments have to be invoked (Achorn and Rosenthal, 2019).

7.7 Social evolution

It can be difficult to assign a single mechanism of natural selection or sexual selection as the sole one driving trait evolution. For example, many traits function in more than one context. They can simultaneously be ornaments and armaments in males (Berglund et al., 1996). Male traits that are used by females in mate choice can also be used in male competition, but more importantly can play a role in competition for other resources, such as food or access to breeding sites. The same is likely true for females. This complexity was realized in a series of papers by West-Eberhard (1979, 1983, 1989), who proposed using the concept of social selection as a more encompassing concept (Lyon and Montgomerie, 2012; Tobias et al., 2012). This framework allows us to consider sexual selection and natural selection in a united way. An interesting test of the predictions made by social selection was provided using African starlings (Sturnidae). These species are breeding cooperatively and are predicted to show less sexual dimorphism. In a comparison of cooperative and noncooperative species, selection acted differently on females and males. This highlights the importance of ornamental traits for female–female interactions (Rubenstein and Lovette, 2009; Rubenstein, 2012a, 2012b). Another study looked at plumage evolution in New World blackbirds and found that differences in plumage are more likely owing to females evolving more drab plumages (Irwin, 1994), not males increasing ornamentation. Yet another study compared monomorphic and dimorphic lekking species and concluded that social selection is likely caused by social competition affecting both sexes (Trail, 1990). A comparative study of plumage evolution in birds in general and of beak coloration in zebra finches also supports this idea (Dale et al., 2015; Dey et al., 2015). Further support comes from a study of dewlaps in *Anolis* lizards (Harrison and Poe, 2012). Overall, I find the support for the role of social evolution in the evolution and maintenance of female ornaments stronger as compared with the genetic correlation hypothesis. It seems possible that genetic correlations are weaker than sometimes assumed and influenced by social traits (Morris et al., 2020).

7.8 Short summary

Ornaments in females are still a bit enigmatic. It seems increasingly doubtful—although still entirely possible—that they are mere byproducts of selection on male ornaments. They might have evolved under selection by male choice in a mechanism parallel to the evolution of male ornaments via female choice, but it seems that social selection is very important in the origin and evolution of female ornaments. If female ornaments evolve under male mate choice, theory predicts that such selection will be more tempered and lead to less costly arguments, because such traits might be too detrimental for females. Mechanisms for the expression of female ornaments involve sex hormones, but the genetic scaffolding for this is poorly known. There

seems to be a strong bias toward visual ornaments, with a little bit known about acoustic communication, but almost nothing on other sensory systems. Likely we are not recognizing many female ornaments because we are not looking for them. Finally, more studies utilizing a phylogenetic framework are needed.

7.9 Further reading

Several papers provide excellent further reading, but the most recent one by Doutrelant et al. stands out (Doutrelant et al., 2020). Other works are by Amundsen (2000) and Nordeide et al. (2013).

7.10 References

ACHORN, A. M. & ROSENTHAL, G. G. 2019. It's not about him: mismeasuring "good genes" in sexual selection. *Trends in Ecology & Evolution*, 35, 206–19.

ADKINS-REGAN, E. 2005. *Hormones and Animal Social Behavior*. Princeton, NJ: Princeton University Press.

AHNESJÖ, I., BREALEY, J. C., GÜNTER, K. P., MARTINOSSI-ALLIBERT, I., MORINAY, J., SILJESTAM, M., STÅNGBERG, J., & VASCONCELOS, P. 2020. Considering gender-biased assumptions in evolutionary biology. *Evolutionary Biology*, 47, 1–5.

ALCOCK, J. 2016. The mating behavior of an undescribed species of *Rhamphomyia* (Diptera: Empididae). *Journal of Insect Behavior*, 29, 153–61.

AMUNDSEN, T. 2000. Why are female birds ornamented? *Trends in Ecology & Evolution*, 15, 149–55.

AMUNDSEN, T. & FORSGREN, E. 2001. Male mate choice selects for female coloration in a fish. *Proceedings of the National Academy of Sciences of the United States of America*, 98, 13155–160.

AMUNDSEN, T. & FORSGREN, E. 2003. Male preference for colourful females affected by male size in a marine fish. *Behavioral Ecology and Sociobiology*, 54, 55–64.

AMUNDSEN, T., FORSGREN, E., & HANSEN, L. T. 1997. On the function of female ornaments: male bluethroats prefer colourful females. *Proceedings of the Royal Society of London—Series B: Biological Sciences*, 264, 1579–86.

ANDERSSON, M. 1994. *Sexual Selection*. Princeton, NJ: Princeton University Press.

ASSIS, B. A., SWIERK, L., & LANGKILDE, T. 2018. Performance, behavior and offspring morphology may offset reproductive costs of male-typical ornamentation for female lizards. *Journal of Zoology*, 306, 235–42.

BADYAEV, A. V. & HILL, G. E. 2003. Avian sexual dichromatism in relation to phylogeny and ecology. *Annual Review of Ecology, Evolution, and Systematics*, 34, 27–49.

BASOLO, A. L. 1990. Female preference predates the evolution of the sword in swordtail fish. *Science,* 250, 808–10.

BASOLO, A. L. 1995. A further examination of a pre-existing bias favouring a sword in the genus *Xiphophorus. Animal Behaviour,* 50, 365–75.

BASOLO, A. L. 2002. Female discrimination against sworded males in a poeciliid fish. *Animal Behaviour,* 63, 463–8.

BASOLO, A. L. & DELANEY, K. J. 2001. Male biases for male characteristics in females in *Priapella olmecae* and *Xiphophorus helleri* (Family Poeciliidae). *Ethology,* [print] 107, 431–8.

BATEMAN, A. J. 1948. Intra-sexual selection in *Drosophila. Heredity,* 2, 349–68.

BENNETT, A., CUTHILL, I., & NORRIS, K. 1994. Sexual selection and the mismeasure of color. *The American Naturalist,* 144, 848–60.

BERGLUND, A., BISAZZA, A., & PILASTRO, A. 1996. Armaments and ornaments: an evolutionary explanation of traits of dual utility. *Biological Journal of the Linnean Society,* 58, 385–99.

BERNET, P., ROSENQVIST, G., & BERGLUND, A. 1998. Female-female competition affects female ornamentation in the sex-role reversed pipefish *Syngnathus typhle. Behaviour,* 135, 535–50.

BERSON, J. D. & SIMMONS, L. W. 2019. Female cuticular hydrocarbons can signal indirect fecundity benefits in an insect. *Evolution,* 73, 982–9.

BLEIWEISS, R. 1992. Reversed plumage ontogeny in a female hummingbird—implications for the evolution of iridescent colors and sexual dichromatism. *Biological Journal of the Linnean Society,* 47, 183–95.

BLEIWEISS, R. 1997. Covariation of sexual dichromatism and plumage colours in lekking and non-lekking birds: a comparative analysis. *Evolutionary Ecology,* 11, 217–35.

BOND, A. L., STANDEN, R. A., DIAMOND, A. W., & HOBSON, K. A. 2016. Sexual size dimorphism and discriminant functions for predicting the sex of Atlantic puffins (*Fratercula arctica*). *Journal of Ornithology,* 157, 875–83.

BURLEY, N. & COOPERSMITH, C. B. 1987. Bill color preferences of zebra finches. *Ethology,* 76, 133–51.

BURLEY, N., KRANTZBERG, G., & RADMAN, P. 1982. Influence of colour-banding on the conspecific preferences of zebra finches. *Animal Behaviour,* 30, 444–55.

BURNS, K. J. 1998. A phylogenetic perspective on the evolution of sexual dichromatism in tanagers (Thraupidae): the role of female versus male plumage. *Evolution,* 52, 1219–24.

BUSS, D. 2015. *Evolutionary Psychology.* Abingdon: Routledge.

CHAPMAN, T., LIDDLE, L. F., KALB, J. M., WOLFNER, M. F., & PARTRIDGE, L. 1995. Cost of mating in *Drosophila melanogaster* females is mediated by male accessory gland products. *Nature,* 373, 241.

CLARK, J. M. 1997. "The ants were duly visited": making sense of John Lubbock, scientific naturalism and the senses of social insects. *The British Journal for the History of Science,* 30, 151–76.

CLUTTON-BROCK, T. 2009. Sexual selection in females. *Animal Behaviour,* 77, 3–11.

CORDERO, A., CARBONE, S. S., & UTZERI, C. 1998. Mating opportunities and mating costs are reduced in androchrome female damselflies, *Ischnura elegans* (Odonata). *Animal Behaviour,* 55, 185–97.

CORNWALLIS, C. K. & BIRKHEAD, T. R. 2007. Experimental evidence that female ornamentation increases the acquisition of sperm and signals fecundity. *Proceedings of the Royal Society of London—Series B: Biological Sciences,* 274, 583–90.

COTTON, S., SMALL, J., HASHIM, R., & POMIANKOWSKI, A. 2010. Eyespan reflects reproductive quality in wild stalk-eyed flies. *Evolutionary Ecology,* 24, 83–95.

COX, R. M., SKELLY, S. L., LEO, A., & JOHN-ALDER, H. B. 2005. Testosterone regulates sexually dimorphic coloration in the eastern fence lizard, *Sceloporus undulatus. Copeia,* 2005, 597–08.

CUADRADO, M. 1998. The influence of female size on the extent and intensity of mate guarding by males in *Chamaeleo chamaeleon. Journal of Zoology,* 246, 351–8.

CUADRADO, M. 2000. Body colors indicate the reproductive status of female common chameleons: experimental evidence for the intersex communication function. *Ethology,* 106, 79–91.

CUERVO, J. J., DELOPE, F., & MOLLER, A. P. 1996. The function of long tails in female barn swallows (*Hirundo rustica*): an experimental study. *Behavioral Ecology,* 7, 132–6.

DALE, J., DEY, C. J., DELHEY, K., KEMPENAERS, B., & VALCU, M. 2015. The effects of life history and sexual selection on male and female plumage colouration. *Nature,* 527, 367–70.

DAWKINS, R. 1982. *The Extended Phenotype.* Oxford: Oxford University Press.

DELHEY, K., DALE, J., VALCU, M., & KEMPENAERS, B. 2019. Reconciling ecogeographical rules: rainfall and temperature predict global colour variation in the largest bird radiation. *Ecology Letters,* 22, 726–36.

DESCHNER, T., HEISTERMANN, M., HODGES, K., & BOESCH, C. 2003. Timing and probability of ovulation in relation to sex skin swelling in wild West African chimpanzees, *Pan troglodytes verus. Animal Behaviour,* 66, 551–60.

DESCHNER, T., HEISTERMANN, M., HODGES, K., & BOESCH, C. 2004. Female sexual swelling size, timing of ovulation, and male behavior in wild West African chimpanzees. *Hormones and behavior,* 46, 204–15.

DEY, C. J., VALCU, M., KEMPENAERS, B., & DALE, J. 2015. Carotenoid-based bill coloration functions as a social, not sexual, signal in songbirds (Aves: Passeriformes). *Journal of Evolutionary Biology,* 28, 250–8.

DIXSON, A. 2015. Primate sexuality. *In:* BLACKWOOD, E. et al. (eds.) *The International Encyclopedia of Human Sexuality.* Hoboken, NJ: John Wiley & Son, pp. 861–1042.

DOMB, L. G. & PAGEL, M. 2001. Sexual swellings advertise female quality in wild baboons. *Nature,* 410, 204.

DOUGLAS, P. H., HOHMANN, G., MURTAGH, R., THIESSEN-BOCK, R., & DESCHNER, T. 2016. Mixed messages: wild female bonobos show high variability in the timing of ovulation in relation to sexual swelling patterns. *BMC Evolutionary Biology,* 16, 140.

DOUTRELANT, C., FARGEVIEILLE, A., & GRÉGOIRE, A. 2020. Evolution of female coloration: what have we learned from birds in general and blue tits in particular. *In:* NAGUIB, M., BARRETT, L., HEALY, S. D., PODOS, J., SIMMONS, L. W., & ZUK, M. (eds.) *Advances in the Study of Behavior.* London: Academic Press.

DOUTRELANT, C., GRÉGOIRE, A., GOMEZ, D., STASZEWSKI, V., ARNOUX, E., TVERAA, T., FAIVRE, B., & BOULINIER, T. 2013. Colouration in Atlantic puffins and blacklegged kittiwakes: monochromatism and links to body condition in both sexes. *Journal of Avian Biology,* 44, 451–60.

DUNNING, J., DIAMOND, A. W., CHRISTMAS, S. E., COLE, E.-L., HOLBERTON, R. L., JACKSON, H. J., KELLY, K. G., BROWN, D., ROJAS RIVERA, I., & HANLEY, D. 2018. Photoluminescence in the bill of the Atlantic puffin *Fratercula arctica. Bird Study,* 65, 570–3.

DUTOUR, M. & RIDLEY, A. R. 2020. Females sing more often and at higher frequencies than males in Australian magpies. *Behavioural Processes,* 172, 104045.

EMERSON, S. B. & BOYD, S. K. 1999. Mating vocalizations of female frogs: control and evolutionary mechanisms. *Brain, Behavior and Evolution,* 53, 187–97.

EMERY THOMPSON, M. & WRANGHAM, R. W. 2008. Male mating interest varies with female fecundity in *Pan troglodytes schweinfurthii* of Kanyawara, Kibale National Park. *International Journal of Primatology,* 29, 885–905.

ENDLER, J. A. & BASOLO, A. L. 1998. Sensory ecology, receiver biases and sexual selection. *Trends in Ecology & Evolution,* 13, 415–20.

FAN, J., LIU, F., WU, J., & DAI, W. 2004. Visual perception of female physical attractiveness. *Proceedings of the Royal Society of London—Series B: Biological Sciences,* 271, 347–52.

FOOTE, C. J., BROWN, G. S., & HAWRYSHYN, C. W. 2004. Female colour and male choice in sockeye salmon: implications for the phenotypic convergence of anadromous and nonanadromous morphs. *Animal Behaviour,* 67, 69–83.

GAZDA, M. A., ARAÚJO, P. M., LOPES, R. J., TOOMEY, M. B., ANDRADE, P., AFONSO, S., MARQUES, C., NUNES, L., PEREIRA, P., TRIGO, S., HILL, G. E., CORBO, J. C., & CARNEIRO, M. 2020. A genetic mechanism for sexual dichromatism in birds. *Science,* 368, 1270–4.

GOSDEN, T. P. & SVENSSON, E. I. 2009. Density-dependent male mating harassment, female resistance, and male mimicry. *American Naturalist,* 173, 709–21.

GRIGGIO, M., DEVIGILI, A., HOI, H., & PILASTRO, A. 2009. Female ornamentation and directional male mate preference in the rock sparrow. *Behavioral Ecology,* 20, 1072–8.

GRIGGIO, M., VALERA, F., CASAS, A., & PILASTRO, A. 2005. Males prefer ornamented females: a field experiment of male choice in the rock sparrow. *Animal Behaviour,* 69, 1243–50.

HAMILTON, W. D. & ZUK, M. 1982. Heritable true fitness and bright birds: a role for parasites. *Science,* 218, 384–7.

HANSSEN, S. A., BUSTNES, J. O., TVERAA, T., HASSELQUIST, D., VARPE, O., & HENDEN, J.-A. 2009. Individual quality and reproductive effort mirrored in white wing plumage in both sexes of south polar skuas. *Behavioral Ecology,* 20, 961–6.

HARRISON, A. & POE, S. 2012. Evolution of an ornament, the dewlap, in females of the lizard genus *Anolis. Biological Journal of the Linnean Society,* 106, 191–201.

HAYES, T., HASTON, K., TSUI, M., HOANG, A., HAEFFELE, C., & VONK, A. 2002. Feminization of male frogs in the wild. *Nature,* 419, 895–6.

HEINSOHN, R. 2008. Ecology and evolution of the enigmatic eclectus parrot (*Eclectus roratus*). *Journal of Avian Medicine and Surgery,* 22, 146–51.

HEINSOHN, R., LEGGE, S., & ENDLER, J. A. 2005. Extreme reversed sexual dichromatism in a bird without sex role reversal. *Science (Washington DC),* 309, 617–9.

HEISTERMANN, M., MÖHLE, U., VERVAECKE, H., VAN ELSACKER, L., & KEITH HODGES, J. 1996. Application of urinary and fecal steroid measurements for monitoring ovarian function and pregnancy in the bonobo (*Pan paniscus*) and evaluation of perineal swelling patterns in relation to endocrine events. *Biology of Reproduction,* 55, 844–53.

HEWS, D. K., KNAPP, R., & MOORE, M. C. 1994. Early exposure to androgens affects adult expression of alternative male types in tree lizards. *Hormones and Behavior,* 28, 96–115.

HILL, G. E. 1993. Male mate choice and the evolution of female plumage coloration in the house finch. *Evolution,* 47, 1515–25.

HILL, G. E. 2015. Sexiness, individual condition, and species identity: the information signaled by ornaments and assessed by choosing females. *Evolutionary Biology,* 42, 251–9.

HÖGLUND, J. 1988. Chorusing behaviour, a density dependent alternative strategy in male common toads (*Bufo bufo*). *Ethology,* 79, 324–32.

HUNTER, F. D. & BUSSIÈRE, L. F. 2019. Comparative evidence supports a role for reproductive allocation in the evolution of female ornament diversity. *Ecological Entomology,* 44, 324–32.

HUTH, H.-H. & BURKHARDT, D. 1972. Der spektrale sehbereich eines violettohr-kolibris. *Naturwissenschaften,* 59, 650–650.

IRWIN, R. E. 1994. The evolution of plumage dichromatism in the new world blackbirds—social selection on female brightness. *American Naturalist,* 144, 890–907.

JOHNSEN, T. S., HENGEVELD, J. D., BLANK, J. L., YASUKAWA, K., & NOLAN, V. 1996. Epaulet brightness and condition in female red-winged blackbirds. *Auk,* 113, 356–62.

JONES, I. L. & HUNTER, F. M. 1993. Mutual sexual selection in a monogamous seabird. *Nature,* 362, 238–39.

JONES, I. L. & HUNTER, F. M. 1999. Experimental evidence for mutual inter- and intrasexual selection favouring a crested auklet ornament. *Animal Behaviour,* 57, 521–8.

JUSTIN MARSHALL, N. 2000. Communication and camouflage with the same "bright" colours in reef fishes. *Philosophical Transactions of the Royal Society of London, Series B—Biological Sciences,* 355, 1243–8.

KELLEY, L. A. & ENDLER, J. A. 2012. Illusions promote mating success in great bowerbirds. *Science,* 335, 335–8.

KNIEL, N., BENDER, S. & WITTE, K. 2016. Sex-specific audience effect in the context of mate choice in zebra finches. *PLoS ONE,* 11, e0147130.

KNIEL, N., DÜRLER, C., HECHT, I., HEINBACH, V., ZIMMERMANN, L., & WITTE, K. 2015a. Novel mate preference through mate-choice copying in zebra finches: sexes differ. *Behavioral Ecology,* 26, 647–55.

KNIEL, N., MÜLLER, K., & WITTE, K. 2017. The role of the model in mate-choice copying in female zebra finches. *Ethology,* 123, 412–8.

KNIEL, N., SCHMITZ, J., & WITTE, K. 2015b. Quality of public information matters in mate-choice copying in female zebra finches. *Frontiers in Zoology,* 12, 26.

KOKKO, H., BOOKSMYTHE, I., & JENNIONS, M. D. 2015. Mate-sampling costs and sexy sons. *Journal of Evolutionary Biology,* 28, 259–66.

KRAAIJEVELD, K. 2014. Reversible trait loss: the genetic architecture of female ornaments. *Annual Review of Ecology, Evolution, and Systematics,* 45, 159–77.

LAMPRECHT, J. 1986. Structure and causation of the dominance hierarchy in a flock of bar-headed geese (*Anser indicus*). *Behaviour,* 96, 28–48.

LANDE, R. 1980. Sexual dimorphism, sexual selection, and adaptation in polygenic characters. *Evolution,* 34, 292–305.

LANGMORE, N. E. 1998. Functions of duet and solo songs of female birds. *Trends in Ecology & Evolution,* 13, 136–40.

LANGMORE, N. E. & BENNETT, A. T. D. 1999. Strategic concealment of sexual identity in an estrildid finch. *Proceedings of the Royal Society of London—Series B: Biological Sciences,* 266, 543–50.

LEBAS, N. R. 2006. Female finery is not for males. *Trends in Ecology & Evolution,* 21, 170–3.

LEBAS, N. R., HOCKHAM, L. R., & RITCHIE, M. G. 2004. Sexual selection in the gift-giving dance fly, *Rhamphomyia sulcata,* favors small males carrying small gifts. *Evolution,* 58, 1763–72.

LEBAS, N. R. & MARSHALL, N. J. 2000. The role of colour in signalling and male choice in the agamid lizard *Ctenophorus ornatus. Proceedings of the Royal Society of London—Series B: Biological Sciences,* 267, 445–52.

LIM, M. L. M. & LI, D. 2006. Extreme ultraviolet sexual dimorphism in jumping spiders (Araneae: Salticidae). *Biological Journal of the Linnean Society,* 89, 397–406.

LINVILLE, S. U., BREITWISCH, R., & SCHILLING, A. J. 1998. Plumage brightness as an indicator of parental care in northern cardinals. *Animal Behaviour,* 55, 119–27.

LUBBOCK, J. 1882. *Ants, Bees, and Wasps: a Record of Observations on the Habits of the Social Hymenoptera.* London: Kegan Paul, Trench.

LOEB, M. L. G. 2003. Evolution of egg dumping in a subsocial insect. *The American Naturalist,* 161, 129–42.

LÜDTKE, D. U. & FOERSTER, K. 2018. Choosy males court both large, colourful females and less colourful but responsive females for longer. *Animal Behaviour,* 146, 1–11.

LÜDTKE, D. U. & FOERSTER, K. 2019. A female color ornament honestly signals fecundity. *Frontiers in Ecology and Evolution,* 7, 432.

LYON, B. E. & MONTGOMERIE, R. 2012. Sexual selection is a form of social selection. *Philosophical Transactions of the Royal Society of London, Series B—Biological Sciences,* 367, 2266–73.

MAFRA, A. L., VARELLA, M. A. C., DEFELIPE, R. P., ANCHIETA, N. M., DE ALMEIDA, C. A. G., & VALENTOVA, J. V. 2020. Makeup usage in women as a tactic to attract mates and compete with rivals. *Personality and Individual Differences,* 163, 110042.

MAKOWICZ, A. M., TANNER, J. C., DUMAS, E., SILER, C. D., & SCHLUPP, I. 2016. Pre-existing biases for swords in mollies (*Poecilia*). *Behavioral Ecology,* 27, 175–84.

MARTIN, T. E. & BADYAEV, A. V. 1996. Sexual dichromatism in birds: importance of nest predation and nest location for females versus males. *Evolution,* 50, 2454–60.

MASSARO, M., DAVIS, L. S., & DARBY, J. T. 2003. Carotenoid-derived ornaments reflect parental quality in male and female yellow-eyed penguins (*Megadyptes antipodes*). *Behavioral Ecology and Sociobiology*, 55, 169–75.

MATSUMOTO, K., SOGABE, A., & YANAGISAWA, Y. 2010. Male ornamentation in a sex-role reversed pipefish *Corythoichthys haematopterus*. *Ethology*, 116, 226–32.

MATSUMOTO, K. & YANAGISAWA, Y. 2001. Monogamy and sex role reversal in the pipefish *Corythoichthys haematopterus*. *Animal Behaviour*, 61, 163–70.

MCCOY, E., SYSKA, N., PLATH, M., SCHLUPP, I., & RIESCH, R. 2011. Mustached males in a tropical poeciliid fish: emerging female preference selects for a novel male trait. *Behavioral Ecology and Sociobiology*, 65, 1437–45.

MCLENNAN, D. A. 1994. Changes in female colour across the ovulatory cycle in the brook stickleback, *Culaea inconstans* (Kirtland). *Canadian Journal of Zoology*, 72, 144–53.

MCLENNAN, D. A. 1995. Male mate choice based upon female nuptial coloration in the brook stickleback, *Culaea inconstans* (Kirtland). *Animal Behaviour*, 50, 213–21.

MÉNDEZ-JANOVITZ, M. & MACÍAS GARCIA, C. 2017. Do male fish prefer them big and colourful? Non-random male courtship effort in a viviparous fish with negligible paternal investment. *Behavioral Ecology and Sociobiology*, 71, 160.

MÉNDEZ-JANOVITZ, M., GONZALEZ-VOYER, A., & MACÍAS GARCIA, C. 2019. Sexually selected sexual selection: can evolutionary retribution explain female ornamental colour? *Journal of Evolutionary Biology*, 32, 833–43.

MØLLER, A. P. 1993. Sexual selection in the barn swallow *Hirundo rustica*. Female tail ornaments. *Evolution*, 47, 417–31.

MONAGHAN, P., METCALFE, N. B., & HOUSTON, D. C. 1996. Male finches selectively pair with fecund females. *Proceedings of the Royal Society of London—Series B: Biological Sciences*, 263, 1183–6.

MORALES, M., GIGENA, D. J., BENITEZ-VIEYRA, S. M., & VALDEZ, D. J. 2020. Subtle sexual plumage color dimorphism and size dimorphism in a South American colonial breeder, the monk parakeet (*Myiopsitta monachus*). *Avian Research*, 11, 1–9.

MORRIS, T. T., DAVIES, N. M., HEMANI, G., & SMITH, G. D. 2020. Population phenomena inflate genetic associations of complex social traits. *Science Advances*, 6, eaay0328.

MORROW, E. H., ARNQVIST, G., & PITNICK, S. 2003. Adaptation versus pleiotropy: why do males harm their mates? *Behavioral Ecology*, 14, 802–06.

MUMA, K. E. & WEATHERHEAD, P. J. 1989. Male traits expressed in females—direct or indirect sexual selection. *Behavioral Ecology and Sociobiology*, 25, 23–31.

MUNZ, T. 2016. *The Dancing Bees: Karl von Frisch and the Discovery of the Honeybee Language.* Chicago,IL: University of Chicago Press.

NORDEIDE, J., RUDOLFSEN, G., & EGELAND, E. 2006. Ornaments or offspring? Female sticklebacks (*Gasterosteus aculeatus* L.) trade off carotenoids between spines and eggs. *Journal of Evolutionary Biology*, 19, 431–9.

NORDEIDE, J. T., KEKALAINEN, J., JANHUNEN, M., & KORTET, R. 2013. Female ornaments revisited—are they correlated with offspring quality? *Journal of Animal Ecology*, 82, 26–38.

NUNN, C. L. 1999. The evolution of exaggerated sexual swellings in primates and the graded-signal hypothesis. *Animal Behaviour*, 58, 229–46.

ODOM, K. J., HALL, M. L., RIEBEL, K., OMLAND, K. E., & LANGMORE, N. E. 2014. Female song is widespread and ancestral in songbirds. *Nature Communications*, 5, 3379.

OLIVEIRA, R. F., TABORSKY, M., & BROCKMANN, H. J. (eds.) 2008. *Alternative Reproductive Tactics.* Cambridge: Cambridge University Press.

OWENS, I. P. F., BURKE, T., & THOMPSON, D. B. A. 1994. Extraordinary sex-roles in the eurasian dotterel—female mating arenas, female-female competition, and female mate choice. *American Naturalist*, 144, 76–100.

PATRICELLI, G. L., UY, J. A. C., & BORGIA, G. 2003. Multiple male traits interact: attractive bower decorations facilitate attractive behavioural displays in satin bowerbirds. *Proceedings of the Royal Society of London—Series B: Biological Sciences*, 270, 2389–95.

PIRES, T. H., BORGHEZAN, E. A., CUNHA, S. L., LEITÃO, R. P., PINTO, K. S., & ZUANON, J. 2019. Sensory drive in colourful waters: morphological variation suggests combined natural and sexual selection in an Amazonian fish. *Biological Journal of the Linnean Society*, 127, 351–60.

PRICE, D. K. & BURLEY, N. T. 1994. Constraints on the evolution of attractive traits: selection in male and female zebra finches. *The American Naturalist*, 144, 908–34.

PRICE, J. J. 2015. Rethinking our assumptions about the evolution of bird song and other sexually dimorphic signals. *Frontiers in Ecology and Evolution*, 3, 40.

PRICE, T. & BIRCH, G. L. 1996. Repeated evolution of sexual color dimorphism in passerine birds. *Auk*, 113, 842–8.

PROKOP, Z. M., MICHALCZYK, L., DROBNIAK, S. M., HERDEGEN, M., & RADWAN, J. 2012. Meta-analysis suggests choosy females get sexy sons more than "good genes." *Evolution*, 66, 2665–73.

PUSEY, A. E. & PACKER, C. 1994. Infanticide in lions: consequences and counterstrategies. *In*: PARMIGIANI, S. & VOM SAAL, F. S. (eds.) *Infanticide and Parental Care.*

Chur, Switzerland, Harwood Academic Publishers, pp. 277–99.

REYER, H.-U., BOLLMANN, K., SCHLAPFER, A. R., SCHYMAINDA, A., & KLECACK, G. 1997. Ecological determinants of extrapair fertilizations and egg dumping in Alpine water pipits (*Anthus spinoletta*). *Behavioral Ecology*, 8, 534–43.

RIEBEL, K. 2003. The "mute" sex revisited: vocal production and perception learning in female songbirds. *Advances in the Study of Behavior*, 33, 49–86.

ROHDE, P. A., JOHNSEN, A., & LIFJELD, J. T. 1999. Female plumage coloration in the bluethroat: no evidence for an indicator of maternal quality. *Condor*, 101, 96–104.

RUBENSTEIN, D. R. 2012a. Family feuds: social competition and sexual conflict in complex societies. *Philosophical Transactions of the Royal Society of London, Series B—Biological Sciences*, 367, 2304–13.

RUBENSTEIN, D. R. 2012b. Sexual and social competition: broadening perspectives by defining female roles. *Philosophical Transactions of the Royal Society B—Biological Sciences*, 367, 2248–52.

RUBENSTEIN, D. R. & LOVETTE, I. J. 2009. Reproductive skew and selection on female ornamentation in social species. *Nature*, 462, 786.

RUTOWSKI, R. L. & KEMP, D. J. 2017. Female iridescent colour ornamentation in a butterfly that displays mutual ornamentation: is it a sexual signal? *Animal Behaviour*, 126, 301–07.

RYAN, M. J. 1990. Sexual selection, sensory systems and sensory exploitation. *Oxford Surveys in Evolutionary Biology*, 7, 157–95.

RYAN, M. J. & KEDDY-HECTOR, A. 1992. Directional patterns of female mate choice and the role of sensory biases. *American Naturalist*, 4–35.

RYAN, M. J. & RAND, A. S. 1998. Evoked vocal response in male tungara frogs: pre-existing biases in male responses? *Animal Behaviour*, 56, 1509–16.

SCHARTL, M., SCHLUPP, I., SCHARTL, A., MEYER, M. K., NANDA, I., SCHMID, M., EPPLEN, J. T., & PARZEFALL, J. 1991. On the stability of dispensable constituents of the eukaryotic genome—stability of coding sequences versus truly hypervariable sequences in a clonal vertebrate, the Amazon molly, *Poecilia formosa*. *Proceedings of the National Academy of Sciences of the United States of America*, 88, 8759–63.

SCHEDINA, I. M., GROTH, D., SCHLUPP, I., & TIEDEMANN, R. 2018. The gonadal transcriptome of the unisexual Amazon molly *Poecilia formosa* in comparison to its sexual ancestors, *Poecilia mexicana* and *Poecilia latipinna*. *BMC Genomics*, 19.

SCHIEMENZ, F. 1924. Über den Farbensinn der Fische. *Journal of Comparative Physiology A: Neuroethology, Sensory, Neural, and Behavioral Physiology*, 1, 175–220.

SCHLUPP, I. 2005. The evolutionary ecology of gynogenesis. *Annual Review of Ecology Evolution and Systematics*, 36, 399–417.

SCHLUPP, I. 2018. Male mate choice in livebearing fishes: an overview. *Current Zoology*, 64, 393–403.

SCHLUPP, I., NANDA, I., DÖBLER, M., LAMATSCH, D. K., EPPLEN, J. T., PARZEFALL, J., SCHMID, M., & SCHARTL, M. 1998. Dispensable and indispensable genes in an ameiotic fish, the Amazon molly *Poecilia formosa*. *Cytogenetics and Cell Genetics*, 80, 193–8.

SCHLUPP, I., PARZEFALL, J., EPPLEN, J. T., NANDA, I., SCHMID, M., & SCHARTL, M. 1992. Pseudomale behaviour and spontaneous masculinization in the all-female teleost *Poecilia formosa* (Teleostei: Poeciliidae). *Behaviour*, 122, 88–104.

SCHLUPP, I., RIESCH, R., TOBLER, M., PLATH, M., PARZEFALL, J., & SCHARTL, M. 2010. A novel, sexually selected trait in poeciliid fishes: female preference for mustache-like, rostral filaments in male *Poecilia sphenops*. *Behavioral Ecology and Sociobiology*, 64, 1849–55.

SILLÉN-TULLBERG, B. & MOLLER, A. P. 1993. The relationship between concealed ovulation and mating systems in anthropoid primates: a phylogenetic analysis. *The American Naturalist*, 141, 1–25.

SKARSTEIN, F. & FOLSTAD, I. 1996. Sexual dichromatism and the immunocompetence handicap: an observational approach using Arctic charr. *Oikos*, 1996, 359–67.

SOGABE, A. & YANAGISAWA, Y. 2007. Sex-role reversal of a monogamous pipefish without higher potential reproductive rate in females. *Proceedings of the Royal Society of London—Series B: Biological Sciences*, 274, 2959–63.

STODDARD, M. C. & PRUM, R. O. 2011. How colorful are birds? Evolution of the avian plumage color gamut. *Behavioral Ecology*, 22, 1042–52.

STUART-FOX, D. & MOUSSALLI, A. 2008. Selection for social signalling drives the evolution of chameleon colour change. *PLoS Biology*, 6, e25.

SVENSSON, E. I. & ABBOTT, J. 2005. Evolutionary dynamics and population biology of a polymorphic insect. *Journal of Evolutionary Biology*, 18, 1503–14.

SWIERK, L. & LANGKILDE, T. 2013. Bearded ladies: females suffer fitness consequences when bearing male traits. *Biology Letters*, 9, 20130644.

SWIERK, L. & LANGKILDE, T. 2019. Fitness costs of mating with preferred females in a scramble mating system. *Behavioral Ecology*, 30, 658–65.

SWIERK, L., MYERS, A., & LANGKILDE, T. 2013. Male mate preference is influenced by both female behaviour and morphology. *Animal Behaviour*, 85, 1451–7.

TABORSKY, M. 1994. Sneakers, satellites, and helpers: parasitic and cooperative behavior in fish reproduction. *Advances in the Study of Behavior*, 23, e100.

TAKAHASHI, D. 2000. Conventional sex roles in an amphidromous rhinogobius goby in which females exhibit nuptial coloration. *Ichthyological Research*, 47, 303–6.

TANAKA, K. D. 2015. A colour to birds and to humans: why is it so different? *Journal of Ornithology*, 156, 433–40.

TELLA, J. L., FORFERO, M. G., DONÁZAR, J. A., & HIRALDO, F. 1997. Is the expression of male traits in female lesser kestrels related to sexual selection? *Ethology*, 103, 72–81.

THORNHILL, R. & GRAMMER, K. 1999. The body and face of woman: one ornament that signals quality? *Evolution and Human Behavior*, 20, 105–20.

TOBIAS, J. A., MONTGOMERIE, R., & LYON, B. E. 2012. The evolution of female ornaments and weaponry: social selection, sexual selection and ecological competition. *Philosophical Transactions of the Royal Society of London, Series B—Biological Sciences*, 367, 2274–93.

TOVÉE, M. J., MAISEY, D. S., EMERY, J. L., & CORNELISSEN, P. L. 1999. Visual cues to female physical attractiveness. *Proceedings of the Royal Society of London—Series B: Biological Sciences*, 266, 211–18.

TRAIL, P. W. 1990. Why should lek-breeders be monomorphic. *Evolution*, 44, 1837–52.

VOROBYEV, M., OSORIO, D., BENNETT, A. T. D., MARSHALL, N. J., & CUTHILL, I. C. 1998. Tetrachromacy, oil droplets and bird plumage colours. *Journal of Comparative Physiology A*, 183, 621–33.

WARREN, W. C., GARCIA-PEREZ, R., XU, S., LAMPERT, K. P., CHALOPIN, D., STOECK, M., LOEWE, L., LU, Y., KUDERNA, L., MINX, P., MONTAGUE, M. J., TOMLINSON, C., HILLIER, L. W., MURPHY, D. N., WANG, J., WANG, Z., MACIAS GARCIA, C., THOMAS, G. C. W., VOLFF, J.-N., FARIAS, F., AKEN, B., WALTER, R. B., PRUITT, K. D., MARQUES-BONET, T., HAHN, M. W., KNEITZ, S., LYNCH, M., & SCHARTL, M. 2018. Clonal polymorphism and high heterozygosity in the celibate genome of the Amazon molly. *Nature Ecology & Evolution*, 2, 669–79.

WATKINS, G. G. 1997. Inter-sexual signalling and the functions of female coloration in the tropidurid lizard *Microlophus occipitalis*. *Animal Behaviour*, 53, 843–52.

WEIGMANN, C. & LAMPRECHT, J. 1991. Intraspecific nest parasitism in bar-headed geese, *Anser indicus*. *Animal Behaviour*, 41, 677–88.

WEISS, S. L. & DUBIN, M. 2018. Male mate choice as differential investment in contest competition is affected by female ornament expression. *Current Zoology*, 64, 335–44.

WEISS, S. L., KENNEDY, E. A., & BERNHARD, J. A. 2009. Female-specific ornamentation predicts offspring quality in the striped plateau lizard, *Sceloporus virgatus*. *Behavioral Ecology*, 20, 1063–71.

WEST-EBERHARD, M. J. 1979. Sexual selection, social competition, and evolution. *Proceedings of the American Philosophical Society*, 123, 222–34.

WEST-EBERHARD, M. J. 1983. Sexual selection, social competition, and speciation. *Quarterly Review of Biology*, 58, 155–83.

WEST-EBERHARD, M. J. 1989. Phenotypic plasticity and the origins of diversity. *Annual Review of Ecology and Systematics*, 20, 249–78.

WHITTINGHAM, L. A., KIRKCONNEL, A., & RATCLIFFE, L. M. 1992. Differences in song and sexual dimorphism between Cuban and North American red-winged blackbirds (*Agelaius phoeniceus*). *Auk*, 109, 928–33.

WHITTINGHAM, L. A., KIRKCONNELL, A., & RATCLIFFE, L. M. 1996. Breeding behavior, social organization and morphology of red-shouldered (*Agelaius assimilis*) and tawny-shouldered (*A. humeralis*) blackbirds. *Condor*, 98, 832–6.

WRIGHT, A. A. 1972. The influence of ultraviolet radiation on the pigeon's color discrimination. *Journal of the Experimental Analysis of Behavior*, 17, 325–37.

ZHU, F., SCHLUPP, I., & TIEDEMANN, R. 2017. Allele-specific expression at the androgen receptor alpha gene in a hybrid unisexual fish, the Amazon molly (*Poecilia formosa*). *Plos One*, 12.

ZINNER, D., ALBERTS, S. C., NUNN, C. L., & ALTMANN, J. 2002. Evolutionary biology (communication arising): significance of primate sexual swellings. *Nature*, 420, 142.

ZINNER, D. P., NUNN, C. L., VAN SCHAIK, C. P., & KAPPELER, P. M. 2004. Sexual selection and exaggerated sexual swellings of female primates. *In*: KAPPELER, P. M. & VAN SCHAIK, C. P. (eds.) *Sexual Selection in Primates: New and Comparative Perspectives*. New York: Cambridge University Press, pp. 71–89.

CHAPTER 8

Female–Female Competition

8.1 Brief overview of the chapter

There is no denying that males compete over access to reproductive opportunities. This competition can be covert or take the form of physical fights, which are easily observed in many species. Males take considerable risks when they fight, presumably because the price—access to reproduction—is so high. But how about females? They also often compete for males, but in less risky ways. They also seem to compete less directly for individual males, but more indirectly for resources that males can provide. Nonetheless, female competition is more important in shaping sexual selection than currently thought.

8.2 Introduction

Competition is a ubiquitous feature of ecology and evolution. It is so deeply engrained in everything that it clearly is one of the major organizing principles of life and one of the biggest forces driving evolution. Organisms compete over all types of resources either directly or indirectly (Hardy and Briffa, 2013).

Competition and one other extremely important driver of evolution, the need to reproduce, meet in sexual selection theory. Darwin firmly established the role of competition in sexual selection, but mostly focused on males competing over access to females. The examples for this are abundant, and there is no question that this is a major force behind male traits that are adaptive in competition. At the same time some of these traits can be under selection via female choice. But that males compete does not rule out that females may also compete for resources, including males. However, because

competition is such an important force, it is sometimes difficult to single out the particular competition over mating partners from the background of general competition for any other resources. In addition to direct competition over mates there is also competition over resources that play a role in reproduction. For example, competition for nesting sites could be considered in the context of sexual selection. However, competition for food, which is also important to reproduce, may not be considered in a sexual selection context, because food is also essential to survival, illustrating how blurry the distinction can be.

Male competition for access to females is the first hallmark of Darwin's theory of sexual selection, in addition to female choice. Immediately accepted by the scientists and public of the Victorian era, it attracted much early attention. After the rise of female choice, however, fewer studies on male competition have emerged. The core principles are well understood, however. Males benefit from excluding other males from access to females. In this type of competition, traits can be adaptive as weapons and simultaneously function as ornaments for female choice (Berglund et al., 1996; Emlen, 2008). Weapon size in males has been linked with sexual selection (Bro-Jørgensen, 2007). For example, in African ungulates in many species—but not all—both sexes have horns. Males use them in fighting, but females too (Packer, 1983). However, fighting in females is both less common and less severe, and their horns might be more specialized as defense weapons against predators.

But why should females compete over males, especially given that they are often able to choose? And *if* they fight, females should be selected not to risk harm, certainly not to the degree males may harm each other in open competition. This would be in line with the traditional view of sex roles. In a 2009 popular science book, Meston and Buss said that female competition has been "the blind spot among generations of researchers" (Meston and Buss, 2009, p. 68)—at least for studies in humans. Actually, scientists—going back to Darwin (1871)—have known for a long time that females do fight. An early example is provided by Tinbergen, who reports of several species of birds with escalated fights (Tinbergen, 1936). But what are the evolutionary consequences of female competition? Does it generate sexual selection analogous to that resulting from male competition? Do ornaments and armaments arise from female competition? In a review paper, Rosvall lists several scenarios (Rosvall, 2011b) for when and why females should compete.

First, males may be rare based on the operational sex ratio (OSR). This has been described in small fishes, gobies, and is indeed associated with a flip in choosiness and competition: early in the breeding season females choose among the more abundant males, but later in the season males become rare and choosy, and females compete (Amundsen, 2018) (Chapter 6). Second, females may compete for high-quality partners. Practically by definition they are rare in the pool of available mating partners. A distinction can be made between males that provide more or better direct benefits, and males that provide better indirect benefits. Finally, female competition may simply be a byproduct—again based on a genetic correlation—of male competition. Stockley and Bro-Jørgensen reviewed the literature on female competition in mammals and found that competition for mating partners and related resources is widespread (Stockley and Bro-Jørgensen, 2011). They noted, however, that a clear distinction between natural and sexual selection was difficult. They also highlighted that the role of female competition for sperm of preferred males needs more exploration. In essence the fact that males can be sperm depleted predicts that females might benefit from being the first female to mate with a male that is preferred by multiple females. In addition to a preference for males with certain traits, this argument predicts a certain sequence for matings as a mechanism for male mate choice. Of course, something similar might be found in male mate choice

where males are predicted to mate with the most attractive females first, and with less attractive females later. It is becoming clearer now that female competition and aggression can have a direct adaptive benefit for females. In tree swallows (*Tachycineta bicolor*), for example, females aggressively compete for resources (Rosvall, 2008). Furthermore, in this particular system, the role of steroid hormones is particularly important (Rosvall et al., 2012; Rosvall and Peterson, 2014; Bentz et al., 2019). This also shows that the standard association of different steroids with typical sex roles needs to be evaluated along with the sex roles *per se* (Lipshutz et al., 2019).

What seems important to me is that the quality of a male can change quickly over time as he is interacting with other females, introducing a more dynamic, temporal aspect into the decision-making matrix. As a male mates, his status changes and he may go from high quality to low quality and back again. Research into the effects of mating history is increasing (deJong et al., 1998; Alavi et al., 2016; Guevara-Fiore and Endler, 2018), but more work would be useful.

Furthermore, based on which male sires the most offspring, the best sequence for males mating with a female will vary. With first male precedence, where a majority of the offspring are sired by the first males to mate with a female, males should prefer to mate with a given female first, while the opposite would be true for last male precedence, where most offspring are sired by the last male to mate (Squires et al., 2015). Last male precedence, for example, might select for mate guarding, to ensure paternity. Of course, assessing when best to mate is a challenge for males, and an opportunity for females to exercise cryptic choice (Firman et al., 2017). Correspondingly, females should compete for high-quality males to be either the first or last to mate with them, depending on sperm precedence. Within a social network, sperm competition and depletion predict for example if and how males and females should respond to information on other individuals mating. If, for example, last male precedence prevails it might be adaptive to let other males mate first, because mating last is beneficial. Furthermore, manipulating social information and trying to deceive males about copulations might be adaptive from a male point of view (Plath et al., 2010).

Consequently, sperm depletion may play an important role for females. Females should avoid mating with males that have little or no sperm. But this is likely to happen to males that are preferred by many females. Several studies show that there may be sperm depletion, but not all studies find that to be the case (Swierk et al., 2015). Female guppies (*Poecilia reticulata*) have been found to avoid recently mated males (Scarponi et al., 2015) and seem to use chemical cues to recognize such males (Scarponi and Godin, 2018). In a moth (*Lobesia botrana*), nonvirgin males produce significantly smaller spermatophores and females prefer virgin males because their spermatophores provide more direct benefits to the female (Muller et al., 2016). Male mice (*Mus musculus*) seem to be able to exercise cryptic choice by adjusting their ejaculate to the number of matings they experience (Edwards and Cameron, 2017). How they might do this, however, is not clear. Altogether this indicates that access to the best sperm may facilitate the evolution of direct female competition for males. In addition, females may compete for resources provided by males. The latter mechanism is easier to detect and document.

Female competition has long been studied, but not often in the context of mate choice, but relative to maternal care and or maternal aggression (Moyer, 1968; Scaia et al., 2018). Sometimes, female competition is quite similar to that of males, such as in the cichlid, *Cichlasoma dimerus*, where males and females show comparable behaviors (Scaia et al., 2018). In *Julidochromis regani*, another cichlid species, females display more mouth fighting than males (Ito et al., 2017). Female superb fairy-wrens, *Malurus cyaneus*, a bird from Australia, show aggression in the context of resource defense and have a more pronounced aggressive response to female intruders in low-quality territories than in higher-quality territories (Cain and Langmore, 2016). Relatedness is also important in competition: for example in a solitary ectoparasitoid wasp, *Eupelmus vuilleti*, egg-laying females show increased aggression toward unrelated females (Mathiron et al., 2018, 2019). Together this indicates that competition—and aggression—in females is as complex and probably as important as in males (Stockley and Bro-Jørgensen, 2011). Benefits of female competition have, for example, been documented in a

wren (*Troglodytes aedon*). Here, females that are more aggressive have more eggs and feed the offspring more (Krieg and Getty, 2020). Interestingly, their partners are also feeding more.

In this context it is quite important to note that competitive interactions are not limited to interactions within just one sex, are not limited to binary interactions, and are also not limited to within-species interactions. As with other aspects, selection pressures shaping competition may fall within sexual selection, but also natural selection. Thus, probably the best way to understand female competition and also connect it with female ornamentation is via the concept of social evolution (Tobias et al., 2012).

8.3 Costs of competition

As in male competition, there are costs associated with female competition. As I argued before, males are risking injury, even death in competition to a higher degree because their likelihood of obtaining a mate (or another mate) is often relatively low. Most males die as virgins without ever having reproduced. The cost vs. benefit relationship for females must look different, and there are few examples of females damaging each other in competition.

In female baboons (*Papio cynocephalus anubis*) high rank in females is associated with several benefits, such as improved infant survival, but there are also costs in the form of a higher probability of miscarriage and reduced fertility (Packer et al., 1995). Also, in baboons, females have been physically injured by other females (MacCormick et al., 2012). In another species, *Papio ursinus*, cost of open aggression is associated with female reproduction (Huchard and Cowlishaw, 2011) (Figure 8.1). Another example is the increased rate of senescence that meerkats (*Suricata suricatta*) experience (Sharp and Clutton-Brock, 2011a). Overall, however, it seems that physical damage to females is less common and less severe as compared with males. Of course, competition need not manifest itself in open fights, and many other forms of competition are known. In a comparative study of several species of baboons (*Papio* spp.), female competition for males was suggested to reduce

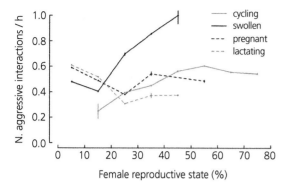

Figure 8.1 Relationship between the rate (number per hour) of aggressive interactions exchanged among females and the mean proportion of females in a given reproductive state in *Papio ursinus* (Huchard and Cowlishaw, 2011) (reproduced from Huchard, E. and Cowlishaw, G. 2011. Female-female aggression around mating: an extra cost of sociality in a multimale primate society. *Behavioral Ecology*, 22, 1003–11.
© 2011. Reprinted by permission of Oxford University Press).

individual birth rate (Dunbar and Sharman, 1983). In tree swallows (*T. bicolor*) more aggressive females incubate less, which translates to a direct cost for females and males (Rosvall, 2010b). In this case, males should probably prefer to mate with less aggressive females. Indeed, males can mitigate the costs of female aggression (Rosvall, 2010a). This illustrates that female competition can be costly to females, maybe even comparable to the cost faced by males in male competition.

8.4 Forms of competition

Competition can come in various shapes and forms. It can be obvious in direct and open competition, like in fighting male deer (with subsequent female choice; McComb, 1991) or toads (Davies and Halliday, 1978), where individuals physically interact, often risking injury or death (Stuart et al., 1987). But it can also be more subtle, and conflicts can be resolved without an escalated fight. Importantly, fighting is governed by the costs and benefits associated with it (Parker, 1974). Proximately, often size is a strong predictor of the outcome of an aggressive encounter, with larger individuals often winning (Franck and Ribowski, 1989). Individuals size each other up and the outcome of a fight becomes so predictable that the smaller individual gives up

without an actual, escalated fight. This can make it difficult to observe competition in action. Nonetheless, we can study the effects resulting from competition. Direct fights and their tactics are best known in males, but females sometimes do engage in such escalated fights, although it seems that the associated cost makes escalated fights with risk of damage or even death less likely and less severe. But clearly, female fighting has been reported in many species, including humans. In mammals, competition between females over breeding resources can be intense and—for example—female ring-tailed lemurs (*Lemur catta*) can take the lead in fights (Jolly and Pride, 1999). In some cooperative breeding mammals, intense fights take place once the breeding female dies (Sharp and Clutton-Brock, 2011b). In this case the stakes are very high, as the follower will be sole reproductive female for some time. This is also found in queenless ants, where, after an egg-laying female worker (a gamergate) dies, intense fighting among the other workers breaks out (Baratte et al., 2006). Interestingly, the competing females are often sisters, apparently violating the general rule that kin should be kind to each other. Related to this, females of spotted hyenas (*Crocuta crocuta*) may engage in siblicide (Hofer and East, 2008), which is also widespread in birds (Mock and Parker, 1997).

In humans (*Homo sapiens*), fights between women may be less common than fights between men, but they are not absent. It is not entirely clear if many fights are over mating partners, but conflicts can turn violent and lead to injuries (Campbell, 1986), although typically less so than conflicts among men (Campbell, 2013). Carrying a weapon, for example, increases the likelihood of involvement in physical altercation in adolescent men, but even more so in women (Lowry et al., 1998).

Individuals can avoid the costs and risks of a full-blown aggressive interaction by issuing credible threats. This is widespread in animals with social rank order systems. In humans, threats of some kind of violence are extremely widespread and play an important role in organizing human society. Threats of punishment are common in our legal system, in many religions, in personal relationships, and—more recently—on social media.

Another form of competition is "mate poaching." This occurs when an individual attempts to break up an existing relationship and to replace one of the partners. This is well known in humans (Parker and Burkley, 2009), and occurs in both genders (Schmitt and Shackelford, 2003).

8.5 What females compete for

8.5.1 Access to mates

We have already seen a few examples of how males may differ in quality, especially sperm. This sets the stage for female competition for high-quality males. Generally, however, female competition for males is not well understood (Rosvall, 2011b; Stockley and Bro-Jørgensen, 2011). Females seem to directly compete for mates mainly in two scenarios. First, in direct competition, they may compete for mates in general, or they can compete for access to high-quality males. The second scenario has a number of interesting implications. If females compete for the best males, how do they, for example, avoid mating with a sperm-depleted male? Sometimes males can be scarce in a population and they may become a limiting resource for females. This shift in OSR is known to lead to male mate choice. This is well studied in a fish, the two-spotted goby (*Gobiusculus flavescens*) (Amundsen, 2018) (Chapter 6). Over the breeding season the OSR changes from male-biased to female-biased. Choosiness was correlated with OSR: early in the season females were choosy, but late in the season, after the OSR had flipped, males were choosy. There was also a corresponding increase in female propensity to show agonistic behavior, while the male propensity went down (Amundsen, 2018). Female competition for access to males may be over access to males *per se*, but especially to high-quality males (Venner et al., 2010). In addition, females may also compete indirectly for resources held by males. An overview of female competition in mammals was provided by Clutton-Brock and Huchard (2013).

Males of some species form coalitions to gain access to mating partners (Van Schaik et al., 2006). One example is groups of males, sometimes closely related, fighting with resident males over a group of females (Packer and Pusey, 1982). This is known in lions (*Panthera leo*) (Chakrabarti and Jhala, 2017), but also many primate species (Freeman et al., 2016; Sicotte et al., 2017; Surbeck et al., 2017), and other species such as dolphins (Connor et al., 2017). Female coalitions, on the other hand, are also well known, sometimes serving as defensive alliances against male aggression (Seyfarth, 1976; Silk et al., 2004; Newton-Fisher, 2006). They can also be a mechanism for social advance (Strauss and Holekamp, 2019), but I did not find reports of female coalitions serving to access males.

8.5.2 Access to reproductive resources

Males are an indispensable resource for female sexual reproduction, but not the only crucial one. In species where males contribute nothing beyond the sperm, females could evolve to select for males that are not sperm depleted. Males, of course, might benefit from hiding sperm depletion from females. One has to be careful, though, because in species with internal fertilization, components of the male ejaculate can be either detrimental, or beneficial to females. In *Drosophila melanogaster*, ejaculates contain substances that reduce female fertility in subsequent matings, benefiting the male at the expense of the female (Chapman et al., 1995; Chapman, 2001), but in the same species a sperm peptide has been found to enhance long-term memory in females (Scheunemann et al., 2019).

We briefly discussed coalitions above, but females can also cooperate with a mating partner to secure important resources. This has been a common explanation for duetting in birds and might be an advantage of a monogamous mating system. Duetting is best studied in birds but is also found in other taxa. The concept has also been expanded beyond song to other signaling modalities (Ręk and Magrath, 2016). In one species of bird, the spectacled parrotlets (*Forpus conspicillatus*), for example, males and females pair up to defend nest sites. These birds are tropical cavity breeders and nesting cavities are rare (Wanker et al., 2005). It seems adaptive for pairs to cooperate and work as a team to secure and defend nesting sites. Such behavior has interesting consequences and the competition between couples appears to also select for the ability of individual recognition (Wanker and

Fischer, 2001). A study on eastern bluebirds (*Sialis sialis*) also suggests that female song is important in individual recognition (Rose et al., 2020). Furthermore, female *Formicarius moniliger*, a songbird from Mexico, show a fine-scaled response to male song playback and seem to respond with territorial defense only to males that are presumably weaker than they are (Kirschel et al., 2020).

An additional dimension of fighting in pairs is illustrated by the interactions shown in acorn woodpeckers (*Melanerpes formicivorus*). These socially breeding birds are limited by trees, in which they store acorns. These trees become hotly contested and lead to openly aggressive interactions between groups of woodpeckers, which include males and females. Interestingly, these fights also provide public information about the involved individual and attract large numbers of uninvolved individuals as an audience (Barve et al., 2020).

While evidence for female competition over resources is relatively common, direct competition for males is not observed often. In moorhens, *Gallinula chloropus*, a widespread Old World waterbird, females compete aggressively for males. Heavier females win more of the fights and associate with males that are heavier (Petrie, 1983). Another study confirmed that male weight influences breeding success via territory size (Gibbons, 1989), pointing toward direct benefits for the females. The fights in females involve jumping up and trying to reach each other with their clawed feet, leading to a relatively high risk of injury and probably also predation because the very conspicuous behavior is likely to attract predators. Interestingly, both males and females in this species are ornamented with a red head plate, and it seems possible that this plate is both ornament and armament (Berglund et al., 1996).

8.5.3 Access to male investment

In many mating systems males make considerable investments in reproduction beyond providing sperm. Given that there is variance in male quality, there might be female preferences for and female competition over the males that provide the most or best resources. We have already discussed this in Chapter 5.

8.5.4 Access to indirect benefits

Female choice for indirect, genetic benefits is generally rare (Achorn and Rosenthal, 2019), but nonetheless there are some well-documented examples for this. If males with "good genes" are rare in a population, it might be beneficial for females to compete for access to these males. For example, if many females converge on mating with a few or even a single male in a lek, that male may be sperm depleted after some time and be a less attractive resource for the females (Chapter 4). Consequently, it might be advantageous to be among the first females to mate with that particular male.

8.6 Infanticide

Infanticide is a behavior where juveniles are killed by adults of their own species. It is a very common behavior found in a large number of taxa, including humans (Hausfater and Hrdy, 2017). Often, but not always, such infanticide is conducted by males. A common example is infanticide by male lions (*Panthera leo*). They can benefit from killing the present offspring of another male as females will reach estrus soon after the killing, allowing the male to mate with the female and sire offspring of their own. This behavior is met with several female counterstrategies (Packer and Pusey, 1983). Importantly, females also kill offspring of other females. This can be viewed as a rather extreme form of competition. The cost to the mother of the killed offspring is very high, and often they resist the infant killing (Agrell et al., 1998). Females are thought to commit infanticide for a number of reasons. If they eat the offspring of other females, this provides them with nutrition, they may acquire access to key resources, or they may eliminate competition for their own offspring (Agrell et al., 1998; Hausfater and Hrdy, 2017).

8.7 Lower-level competition: threats

Reproductive suppression and infanticide may be fairly extreme forms of competition, but more generally, threats can have similar effects. They can also have negative effects on the health, immune status, or stress level in subordinate females. This may cause them to leave a group, often facing high costs for this, even including death (Clutton-Brock and Parker, 1995). A series of studies in fishes showed the cost of female competition between and within species. Size was a very important factor, but not the only one (Makowicz and Schlupp, 2015b; Makowicz et al., 2018a, 2018b, 2020). In primates, threats can be acoustic and visual; in the fish studies cited above the threats were apparently visual. This highlights the complexity of female interactions.

8.8 The role of kinship

While competition is universal, not all competitors should be treated equally in the face of competition. Hamilton's theory of kin selection predicts that close kin should be less competitive with each other than distant relatives or nonkin. However, when important resources can only be gained at the expense of kin, tradeoffs show up and lead to a reduced role of kinship. This was found in a pathogenic bacterium, *Pseudomonas aeruginosa*, where kinship plays an important role in the evolution of iron-scavenging siderophores. In high-competition environments the role of kinship is reduced (Griffin et al., 2004).

The role of kinship was also studied in a system with extreme kinship. In Amazon mollies (*Poecilia formosa*), a clonal fish, relatedness within clones is close to 1, just like we would expect with full twins in other animals. Another clone may still be highly related, but somewhat less than 1. This is different from sexual organisms where relatedness with their own offspring is typically 0.5. Based on this situation one can generally predict that Amazon mollies should be less aggressive with full clonal sisters, more aggressive with other clones, and most aggressive with sexual females. Although the situation is a little more complex, this is what is empirically supported. Interestingly, such fine-tuned aggression requires that Amazon mollies can distinguish between full clonal sisters and others, which they apparently do based on several sensory systems, including vision (Makowicz et al., 2016, 2018a). In another system with clonal kinship, polyembryonic wasp (*Copidosoma floridanum*), relatedness was a better predictor of aggression than resource competition (Giron et al., 2004).

8.9 Reproductive suppression of other females

Maybe the most radical way for females to compete with each other is to prevent other females from having offspring, causing reproductive skew. Such reproductive suppression is a key feature of social animals. It is a hallmark of eusociality (Ross and Keller, 1995), where one or a few individuals reproduce, but not others (Lukas and Clutton-Brock, 2018). In eusocial insects, typically one queen and very few males reproduce, and all other individuals do not. This has interesting implications for female choice (Baer, 2014) and probably limited opportunities for male choice to evolve. Haplodiploidy as known from eusocial hymenoptera, provides an elegant kin selection-based mechanism for the evolution and maintenance of eusociality, but similar mating systems have also evolved in systems without haplodiploidy, such as in termites and naked mole rats. In all cases there seems to be considerable conflict over opportunities for reproduction and often reproductive females seem to suppress other females so that they do not reproduce. In social insects one mechanism for this is via hormones emitted by the queen (Shimoji et al., 2017). In termites, a candidate gene for this, *Neofem2*, has been identified. Silencing this gene triggers worker females, which are otherwise reproductively competent, to behave as if the colony were queenless (Korb et al., 2009). These hormones, however, may also indicate the fertility of the dominant females (Oi et al., 2015).

Not only insects have evolved eusociality but mammals, too. One of the best understood examples is naked mole rats (*Heterocephalus glaber*) (Faulkes and Bennett, 2001; Lukas and Clutton-Brock, 2018), one of the few eusocial mammals. In mammals, such reproductive suppression is also found in several cooperative breeders that are not eusocial. In Kalahari meerkats (*Suricata suricatta*), for example, the degree of suppression seems to depend on the costs and benefits of reproduction of subordinates to the dominant females (Clutton-Brock et al., 2010) with many factors such as relatedness, group size, and the reproductive status of the dominant female. The suppression, when it happens, takes very physical forms, such as eviction of females and killing their offspring (Clutton-Brock et al., 2010). Other examples include marmoset monkeys (*Callithrix jacchus*) (Barrett et al., 1990), where suppression works via pheromones. Other examples include canids (Moehlman and Hofer, 1997), and many more (Wasser and Barash, 1983). Overall, this intensive female competition does not seem to interact much with competition for males, or reproductive resources. Reproductive suppression is also found in males, of course, for example in male marmots (*Marmota marmota*) (Arnold and Dittami, 1997).

8.10 Within- and between-species competition?

Typically, competition for access to males is a within-species affair, but competition for other, more general resources, like food or shelter can also be between species. This is especially clear in species that breed in spaces also used by other species. The resulting conflicts can be resolved in many ways, from sharing a resource—even to the level of parasitism—to aggressively fighting over it. An example is cavity nesting in birds and insects. Competition can be within species, but also interspecific. For example, in tree swallows (*T. bicolor*) (Leffelaar and Robertson, 1985) females compete very aggressively for nesting sites. As is often the case, cavity nesters rely on other species to build the structures they use. The edible dormouse in Eurasia (*Glis glis*), for example, uses naturally occurring cavities, bird-made holes, and also nestboxes, and sometimes competes with birds (Marteau and Sarà, 2015). An example of shared cavities is the co-existence of prairie dogs (*Cynomys ludovicianus*) and burrowing owls (*Speotyto cunicularia*) in the plains of North America, where the prairie dogs serve as hosts for the owls (Desmond et al., 1995; Lantz et al., 2007). In some of these systems a mutualistic relationship arises, while others are competitive. In all of these cases females may compete over breeding sites as a critical resource for their reproduction.

8.11 Armaments in females

In males, sexually selected traits can serve multiple functions. They can be sexually selected ornaments with females paying attention to them in female choice. The same trait may also serve males in male competition as an armament (Berglund et al., 1996). This has been shown, for example, in white-tailed deer (*Odoceilus virginianus*), where females preferred males with larger antlers independent of age (Morina et al., 2018). Of course, armaments and ornaments can be connected and often not only have multiple functions, but also evolve into each other. For example, in fishes of the genus *Poecilia* and *Limia*, male traits that are now considered courtship evolved first in the context of male competition (Goldberg et al., 2019). The same might be true for the sword in green swordtails, *Xiphophorus hellerii*, which have an extension of their tail fin—the eponymous sword. The sword adds length to the male body and size, which is decisive in male agonistic interactions (Franck and Ribowski, 1989, 1993). The sword is also important in female choice, where longer swords are often preferred (Basolo, 1990; Johnson and Basolo, 2004), probably in interaction with other traits (Rosenthal et al., 1996), but not always (Wong and Rosenthal, 2006). Clearly, size can be also important in female competition, where for example in some social species the size of the dominant female allows her to evict other females owing to physical dominance. This is also relevant in light of an argument we made earlier: if females are selected to be larger because female size is beneficial in female competition, this provides an alternative explanation for the widespread male preference for larger females. There might be more to this preference than a fecundity benefit.

One important question in this context is—assuming they do compete for males—why do females have relatively few armaments (Berglund, 2013)? There is some debate over the dual function of traits that are used both in competition and choice (Berglund et al., 1996; Rubenstein, 2012), and sometimes even the distinction seems difficult (Amundsen, 2000). There are some taxa with armaments, such as some African antelopes (Packer, 1983), but overall, if female competition is important, not only ornaments, but also armaments should be found. This is clearly another area for future research. Alternatively, of course, a genetic correlation might also be responsible for the expression of armaments in females, just as it might be for ornaments.

I am providing only a few examples here: in a species of dung beetle, *Onthophagus sagittarius*, females compete for resources. Horns, which are female-specific, are considered to be weapons, and—controlling for size—females with bigger horns had higher reproductive fitness than females with smaller horns. It is interesting that all females experience reduced fitness as compared with a situation without competition, and that smaller females suffer the most (Watson and Simmons, 2010). In some decapod crabs (*Aegla* sp.) both males and females fight within their sexes. Interestingly, the absolute size of their weapons was larger in males, but—at least in one species—not the relative size (Dalosto et al., 2019). This seems to be in agreement with the notion that armaments are relevant to female competition, but that the evolution of them—just as discussed for ornaments—seems influenced by the avoidance of physical harm to females.

8.12 Eavesdropping and audience effects in females, but not in males?

Fighting provides social information that bystanders can use. Females are known to observe male interactions and use the public information provided by the males in future interactions with the males. Often, they subsequently prefer the winner of a fight as a mating partner (Danchin et al., 2004). In most studies on social information females are watching males interact. There are a few examples of male responses to social information. One of them looked at the response of males to female Japanese quail (*Coturnix japonica*) mating (White and Galef Jr., 1999) and found that males avoid places where they have witnessed matings, and another one found deceptive behavior in male fishes (*Poecilia mexicana*). These males led observing males away from their actually preferred female to a less attractive alternative (Plath et al., 2010). In guppies (*Poecilia reticulata*), this deceptive behavior was not found (Makowicz et al., 2010a), although they also show some response to the audience.

In the all-female and clonal Amazon molly (*P. formosa*) male mate choice is very important. In this mating complex, Amazon mollies need to obtain sperm from males of closely related species to trigger embryogenesis. This sperm-dependent parthenogenesis is called gynogenesis. The male DNA does not contribute to the offspring, and males should be under selection to prefer their conspecific females. Consequently, males of the donor species are in a position to choose between their conspecific females and Amazon mollies. But would they be susceptible to audience effects? In a series of experiments Makowicz and colleagues addressed this question and found overwhelming evidence for adjustment of male mate choice depending on the audience present (Makowicz et al., 2010b, 2020; Riesch et al., 2012). At the same time it seems that the clonal females have evolved a very sensitive mechanism to recognize other clones of the Amazon molly (Makowicz and Schlupp, 2015a), and compete with other clonal and sexual females (Makowicz and Schlupp, 2015a; Makowicz et al., 2018a), while the males of the host species show no recognition of the different clones (Makowicz et al., 2018b). As we study more systems for male mate choice, including studies on the social environment and audience effects will be very important. Male mate choice and female competition is likely also influenced by the social conditions, just like male competition and female choice.

8.13 What does it all mean?

Male competition is extremely visible and very well understood. Its role in sexual selection and how it is intertwined with female choice to lead to ornaments and armaments in males is also fairly well understood. However, on female competition, our understanding is rather limited. As with ornamentation in females we seem to be just waking up to their existence and fully understanding their evolution and function will take us a while. How female competition evolved is still a fairly open question. Although I think that a genetic correlation is unlikely, I also think it cannot be ruled out. That females fight less costly than males is intuitive, but much more work on this is needed. Somewhat unsurprisingly, females not only compete directly for mates, but maybe more so for resources provided by males. Overall, female competition is clearly present, important, and adaptive. But many details require much more attention.

8.14 Short summary

Competition is a hallmark of many interactions. Male aggression is very well studied and understood, but female aggression has not received as much interest, yet it is clearly important in interactions between females. They may compete directly for desirable males, or resources they require to reproduce. While likely also very important, female competition may not reach the level of male competition and have less deadly outcomes.

8.15 Further reading

There are several papers that can be used to gain entrance to the fascinating field of female competition and the connection with female ornamentation. A general overview is provided by Amundsen (2018). A careful discussion can be found in Rosvall (2011a, 2011b).

8.16 References

ACHORN, A. M. & ROSENTHAL, G. G. 2019. It's not about him: mismeasuring "good genes" in sexual selection. *Trends in Ecology & Evolution*, 35, 206–19.

AGRELL, J., WOLFF, J. O., & YLÖNEN, H. 1998. Counter-strategies to infanticide in mammals: costs and consequences. *Oikos*, 83, 507–17.

ALAVI, Y., ELGAR, M. A., & JONES, T. M. 2016. Male mating success and the effect of mating history on ejaculate traits in a facultatively parthenogenic insect (*Extatosoma tiaratum*). *Ethology*, 122, 523–30.

AMUNDSEN, T. 2000. Female ornaments: genetically correlated or sexually selected. *In*: ESPMARK, Y., AMUNDSEN, T., & ROSENQVIST, G. (eds.) *Animal Signals: Signalling and Signal Design in Animal Communication*. Trondheim, Norway: Tapir Academic Press, pp. 133–54.

AMUNDSEN, T. 2018. Sex roles and sexual selection: lessons from a dynamic model system. *Current Zoology*, 64, 363–92.

ARNOLD, W. & DITTAMI, J. 1997. Reproductive suppression in male alpine marmots. *Animal Behaviour*, 53, 53–66.

BAER, B. 2014. Sexual selection in social insects. In: SHUKER, D. M. & SIMMONS, L. W. (eds) *The Evolution of Insect Mating Systems*. Oxford: Oxford University Press.

BARATTE, S., COBB, M., & PEETERS, C. 2006. Reproductive conflicts and mutilation in queenless *Diacamma* ants. *Animal Behaviour*, 72, 305–11.

BARRETT, J., ABBOTT, D., & GEORGE, L. 1990. Extension of reproductive suppression by pheromonal cues in subordinate female marmoset monkeys, *Callithrix jacchus*. *Reproduction*, 90, 411–18.

BARVE, S., LAHEY, A. S., BRUNNER, R. M., KOENIG, W. D., & WALTERS, E. L. 2020. Tracking the warriors and spectators of acorn woodpecker wars. *Current Biology*, 30, R982–3.

BASOLO, A. L. 1990. Female preference for male sword length in the green swordtail, *Xiphophorus helleri* (Pisces: Poeciliidae). *Animal Behavior*, 40, 332–38.

BENTZ, A. B., DOSSEY, E. K., & ROSVALL, K. A. 2019. Tissue-specific gene regulation corresponds with seasonal plasticity in female testosterone. *General and Comparative Endocrinology*, 270, 26–34.

BERGLUND, A. 2013. Why are sexually selected weapons almost absent in females? *Current Zoology*, 59, 564–8.

BERGLUND, A., BISAZZA, A., & PILASTRO, A. 1996. Armaments and ornaments: an evolutionary explanation of traits of dual utility. *Biological Journal of the Linnean Society*, 58, 385–99.

BRO-JØRGENSEN, J. 2007. The intensity of sexual selection predicts weapon size in male bovids. *Evolution: International Journal of Organic Evolution*, 61, 1316–26.

CAIN, K. E. & LANGMORE, N. E. 2016. Female song and aggression show contrasting relationships to reproductive success when habitat quality differs. *Behavioral Ecology and Sociobiology*, 70, 1867–77.

CAMPBELL, A. 1986. Self-report of fighting by females: a preliminary study. *The British Journal of Criminology*, 26, 28–46.

CAMPBELL, A. 2013. The evolutionary psychology of women's aggression. *Philosophical Transactions of the Royal Society B—Biological Sciences*, 368, 20130078.

CHAKRABARTI, S. & JHALA, Y. V. 2017. Selfish partners: resource partitioning in male coalitions of Asiatic lions. *Behavioral Ecology*, 28, 1532–9.

CHAPMAN, T. 2001. Seminal fluid-mediated fitness traits in *Drosophila*. *Heredity*, 87, 511–21.

CHAPMAN, T., LIDDLE, L. F., KALB, J. M., WOLFNER, M. F., & PARTRIDGE, L. 1995. Cost of mating in *Drosophila melanogaster* females is mediated by male accessory gland products. *Nature*, 373, 241.

CLUTTON-BROCK, T. H., HODGE, S. J., FLOWER, T. P., SPONG, G. F., & YOUNG, A. J. 2010. Adaptive suppression of subordinate reproduction in cooperative mammals. *American Naturalist*, 176, 664–73.

CLUTTON-BROCK, T. H. & PARKER, G. A. 1995. Punishment in animal societies. *Nature,* 373, 209.

CLUTTON-BROCK, T. & HUCHARD, E. 2013. Social competition and its consequences in female mammals. *Journal of Zoology,* 289, 151–71.

CONNOR, R. C., CIOFFI, W. R., RANDIĆ, S., ALLEN, S. J., WATSON-CAPPS, J., & KRÜTZEN, M. 2017. Male alliance behaviour and mating access varies with habitat in a dolphin social network. *Scientific Reports,* 7, 46354.

DALOSTO, M. M., AYRES-PERES, L., ARAUJO, P. B., SANTOS, S., & PALAORO, A. V. 2019. Pay attention to the ladies: female aggressive behavior and weapon allometry provide clues for sexual selection in freshwater anomurans (Decapoda: Aeglidae). *Behavioral Ecology and Sociobiology,* 73, 127.

DANCHIN, E., GIRALDEAU, L.-A., VALONE, T. J., & WAGNER, R. H. 2004. Public information: from nosy neighbors to cultural evolution. *Science,* 305, 487–91.

DARWIN, C. 1871. *The Descent of Man, and Selection in Relation to Sex.* London: John Murray.

DAVIES, N. B. & HALLIDAY, T. R. 1978. Deep croaks and fighting assessment in toads *Bufo bufo. Nature,* 274, 683.

DEJONG, P. W., BRAKEFIELD, P. M., & GEERINCK, B. P. 1998. The effect of female mating history on sperm precedence in the two-spot ladybird, *Adalia bipunctata* (Coleoptera, Coccinellidae). *Behavioral Ecology,* 9, 559–65.

DESMOND, M. J., SAVIDGE, J. A., & SEIBERT, T. F. 1995. Spatial patterns of burrowing owl (*Speotyto cunicularia*) nests within black-tailed prairie dog (*Cynomys ludovicianus*) towns. *Canadian Journal of Zoology,* 73, 1375–9.

DUNBAR, R. I. M. & SHARMAN, M. 1983. Female competition for access to males affects birth-rate in baboons. *Behavioral Ecology and Sociobiology,* 13, 157–9.

EDWARDS, A. M. & CAMERON, E. Z. 2017. Cryptic male choice: experimental evidence of sperm sex ratio and seminal fluid adjustment in relation to coital rate. *Reproduction Fertility and Development,* 29, 1401–4.

EMLEN, D. J. 2008. The evolution of animal weapons. *Annual Review of Ecology, Evolution, and Systematics,* 39, 387–413.

FAULKES, C. G. & BENNETT, N. C. 2001. Family values: group dynamics and social control of reproduction in African mole-rats. *Trends in Ecology & Evolution,* 16, 184–90.

FIRMAN, R. C., GASPARINI, C., MANIER, M. K., & PIZZARI, T. 2017. Postmating female control: 20 years of cryptic female choice. *Trends in Ecology & Evolution,* 32, 368–82.

FRANCK, D. & RIBOWSKI, A. 1989. Escalating fights for rank-order position between male swordtails (*Xiphophorus helleri*): effects of prior rank-order experience and information transfer. *Behavioral Ecology & Sociobiology,* 24, 133–44.

FRANCK, D. & RIBOWSKI, A. 1993. Dominance hierarchies of male green swordtails (*Xiphophorus helleri*) in nature. *Journal of Fish Biology,* 43, 497–9.

FREEMAN, N. J., YOUNG, C., BARRETT, L., & HENZI, S. P. 2016. Coalition formation by male vervet monkeys (*Chlorocebus pygerythrus*) in South Africa. *Ethology,* 122, 45–52.

GIBBONS, D. W. 1989. Seasonal reproductive success of the moorhen *Gallinula chloropus*: the importance of male weight. *Ibis,* 131, 57–68.

GIRON, D., DUNN, D. W., HARDY, I. C., & STRAND, M. R. 2004. Aggression by polyembryonic wasp soldiers correlates with kinship but not resource competition. *Nature,* 430, 676.

GOLDBERG, D. L., LANDY, J. A., TRAVIS, J., SPRINGER, M. S., & REZNICK, D. N. 2019. In love and war: the morphometric and phylogenetic basis of ornamentation, and the evolution of male display behavior, in the livebearer genus *Poecilia. Evolution,* 73, 360–77.

GRIFFIN, A. S., WEST, S. A., & BUCKLING, A. 2004. Cooperation and competition in pathogenic bacteria. *Nature,* 430, 1024.

GUEVARA-FIORE, P. & ENDLER, J. A. 2018. Female receptivity affects subsequent mating effort and mate choice in male guppies. *Animal Behaviour,* 140, 73–9.

HARDY, I. C. & BRIFFA, M. 2013. *Animal Contests.* Cambridge: Cambridge University Press.

HAUSFATER, G. & HRDY, S. B. 2017. *Infanticide: Comparative and Evolutionary Perspectives.* Abingdon: Routledge.

HOFER, H. & EAST, M. L. 2008. Siblicide in Serengeti spotted hyenas: a long-term study of maternal input and cub survival. *Behavioral Ecology and Sociobiology,* 62, 341–51.

HUCHARD, E. & COWLISHAW, G. 2011. Female-female aggression around mating: an extra cost of sociality in a multimale primate society. *Behavioral Ecology,* 22, 1003–11.

ITO, M. H., YAMAGUCHI, M., & KUTSUKAKE, N. 2017. Sex differences in intrasexual aggression among sex-role-reversed, cooperatively breeding cichlid fish *Julidochromis regani. Journal of Ethology,* 35, 137–44.

JOHNSON, J. B. & BASOLO, A. L. 2004. Predator exposure alters female mate choice in the green swordtail. *Behavioural Ecology,* 14, 619–25.

JOLLY, A. & PRIDE, E. 1999. Troop histories and range inertia of *Lemur catta* at Berenty, Madagascar: a 33-year perspective. *International Journal of Primatology,* 20, 359–73.

KIRSCHEL, A. N., ZANTI, Z., HARLOW, Z. T., VALLEJO, E. E., CODY, M. L., & TAYLOR, C. E. 2020. Females don't always sing in response to male song, but when they do, they sing to males with higher-pitched songs. *Animal Behaviour,* 166, 129–38.

KORB, J., WEIL, T., HOFFMANN, K., FOSTER, K. R., & REHLI, M. 2009. A gene necessary for reproductive suppression in termites. *Science*, 324, 758–8.

KRIEG, C. A. & GETTY, T. 2020. Fitness benefits to intrasexual aggression in female house wrens, *Troglodytes aedon*. *Animal Behaviour*, 160, 79–90.

LANTZ, S. J., CONWAY, C. J., & ANDERSON, S. H. 2007. Multiscale habitat selection by burrowing owls in black-tailed prairie dog colonies. *The Journal of Wildlife Management*, 71, 2664–72.

LEFFELAAR, D. & ROBERTSON, R. J. 1985. Nest usurpation and female competition for breeding opportunities by tree swallows. *The Wilson Bulletin*, 1985, 221–4.

LIPSHUTZ, S., GEORGE, E., BENTZ, A., & ROSVALL, K. 2019. Evaluating testosterone as a phenotypic integrator: from tissues to individuals to species. *Molecular and cellular endocrinology*, 496, 110531.

LOWRY, R., POWELL, K. E., KANN, L., COLLINS, J. L., & KOLBE, L. J. 1998. Weapon-carrying, physical fighting, and fight-related injury among US adolescents. *American Journal of Preventive Medicine*, 14, 122–9.

LUKAS, D. & CLUTTON-BROCK, T. 2018. Social complexity and kinship in animal societies. *Ecology Letters*, 21, 1129–34.

MACCORMICK, H. A., MACNULTY, D. R., BOSACKER, A. L., LEHMAN, C., BAILEY, A., ANTHONY COLLINS, D., & PACKER, C. 2012. Male and female aggression: lessons from sex, rank, age, and injury in olive baboons. *Behavioral Ecology*, 23, 684–91.

MAKOWICZ, A. & SCHLUPP, I. 2015a. Kin recognition in an asexual fish, *Poecilia formosa*. *Integrative and Comparative Biology*, 55, E118.

MAKOWICZ, A. M. & SCHLUPP, I. 2015b. Effects of female-female aggression in a sexual/unisexual species complex. *Ethology*, 121, 903–14.

MAKOWICZ, A. M., MOORE, T., & SCHLUPP, I. 2018a. Clonal fish are more aggressive to distant relatives in a low resource environment. *Behaviour*, 155, 351–67.

MAKOWICZ, A. M., MURRAY, L., & SCHLUPP, I. 2020. Size, species and audience type influence heterospecific female-female competition. *Animal Behaviour*, 159, 47–58.

MAKOWICZ, A. M., MUTHURAJAH, D. S., & SCHLUPP, I. 2018b. Host species of a sexual-parasite do not differentiate between clones of Amazon mollies. *Behavioral Ecology*, 29, 387–92.

MAKOWICZ, A. M., PLATH, M., & SCHLUPP, I. 2010a. Male guppies (*Poecilia reticulata*) adjust their mate choice behaviour to the presence of an audience. *Behaviour*, 147, 1657–74.

MAKOWICZ, A. M., PLATH, M., & SCHLUPP, I. 2010b. Using video playback to study the effect of an audience on male mating behavior in the sailfin molly (*Poecilia latipinna*). *Behavioural Processes*, 85, 36–41.

MAKOWICZ, A. M., TIEDEMANN, R., STEELE, R. N., & SCHLUPP, I. 2016. Kin recognition in a clonal fish, *Poecilia formosa*. *PLoS ONE*, 11.

MARTEAU, M. & SARÀ, M. 2015. Habitat preferences of edible dormouse, *Glis glis italicus*: implications for the management of arboreal mammals in Mediterranean forests. *Folia Zoologica*, 64, 136–51.

MATHIRON, A. G., POTTIER, P., & GOUBAULT, M. 2018. Let the most motivated win: resource value components affect contest outcome in a parasitoid wasp. *Behavioral Ecology*, 29, 1088–95.

MATHIRON, A. G., POTTIER, P., & GOUBAULT, M. 2019. Keep calm, we know each other: kin recognition affects aggressiveness and conflict resolution in a solitary parasitoid. *Animal Behaviour*, 151, 103–11.

MCCOMB, K. E. 1991. Female choice for high roaring rates in red deer, *Cervus elaphus*. *Animal Behaviour*, 41, 79–88.

MESTON, C. M. & BUSS, D. M. 2009. *Why Women Have Sex: Understanding Sexual Motivations from Adventure to Revenge (and Everything in Between)*. London: Macmillan.

MOCK, D. W. & PARKER, G. A. 1997. *The Evolution of Sibling Rivalry*. Oxford: Oxford University Press.

MOEHLMAN, P. D. & HOFER, H. 1997. Cooperative breeding, reproductive suppression, and body mass in canids. *In*: SOLOMAN, N. G. & FRENCH, J. A. (eds.) *Cooperative Breeding in Mammals*. Cambridge: Cambridge University Press, pp. 76–128.

MORINA, D. L., DEMARAIS, S., STRICKLAND, B. K., & LARSON, J. E. 2018. While males fight, females choose: male phenotypic quality informs female mate choice in mammals. *Animal Behaviour*, 138, 69–74.

MOYER, K. E. 1968. Kinds of aggression and their physiological basis. *Communications in Behavioral Biology*, 2, 65–87.

MULLER, K., ARENAS, L., THIERY, D., & MOREAU, J. 2016. Direct benefits from choosing a virgin male in the European grapevine moth, *Lobesia botrana*. *Animal Behaviour*, 114, 165–172.

NEWTON-FISHER, N. E. 2006. Female coalitions against male aggression in wild chimpanzees of the Budongo Forest. *International Journal of Primatology*, 27, 1589–99.

OI, C. A., VAN OYSTAEYEN, A., CALIARI OLIVEIRA, R., MILLAR, J. G., VERSTREPEN, K. J., VAN ZWEDEN, J. S., & WENSELEERS, T. 2015. Dual effect of wasp queen pheromone in regulating insect sociality. *Current Biology*, 25, 1638–40.

PACKER, C. 1983. Sexual dimorphism: the horns of African antelopes. *Science*, 221, 1191–93.

PACKER, C., COLLINS, D., SINDIMWO, A., & GOODALL, J. 1995. Reproductive constraints on aggressive competition in female baboons. *Nature*, 373, 60.

PACKER, C. & PUSEY, A. E. 1982. Cooperation and competition within coalitions of male lions: kin selection or game theory? *Nature*, 296, 740.

PACKER, C. & PUSEY, A. E. 1983. Adaptations of female lions to infanticide by incoming males. *The American Naturalist*, 121, 716–28.

PARKER, G. A. 1974. Assessment strategy and the evolution of fighting behaviour. *Journal of Theoretical Biology*, 47, 223–43.

PARKER, J. & BURKLEY, M. 2009. Who's chasing whom? The impact of gender and relationship status on mate poaching. *Journal of Experimental Social Psychology*, 45, 1016–19.

PETRIE, M. 1983. Female moorhens compete for small fat males. *Science*, 220, 413–15.

PLATH, M., RICHTER, S., SCHLUPP, I., & TIEDEMANN, R. 2010. Misleading mollies: surface- but not cave-dwelling *Poecilia mexicana* males deceive competitors about mating preferences. *Acta Ethologica*, 13, 49–56.

RĘK, P. & MAGRATH, R. D. 2016. Multimodal duetting in magpie-larks: how do vocal and visual components contribute to a cooperative signal's function? *Animal Behaviour*, 117, 35–42.

RIESCH, R., PLATH, M., MAKOWICZ, A. M., & SCHLUPP, I. 2012. Behavioural and life-history regulation in a unisexual/bisexual mating system: does male mate choice affect female reproductive life histories? *Biological Journal of the Linnean Society*, 106, 598–606.

ROSE, E. M., COSS, D. A., HAINES, C. D., DANQUAH, S. A., HILL, R., LOHR, B., & OMLAND, K. E. 2020. Female song in eastern bluebirds varies in acoustic structure according to social context. *Behavioral Ecology and Sociobiology*, 74, 45.

ROSENTHAL, G. G., EVANS, C. S., & MILLER, W. L. 1996. Female preference for dynamic traits in the green swordtail, *Xiphophorus helleri*. *Animal Behaviour*, 51, 811–20.

ROSS, K. G. & KELLER, L. 1995. Ecology and evolution of social organization: insights from fire ants and other highly eusocial insects. *Annual Review of Ecology and Systematics*, 26, 631–56.

ROSVALL, K. A. 2008. Sexual selection on aggressiveness in females: evidence from an experimental test with tree swallows. *Animal Behaviour*, 75, 1603–10.

ROSVALL, K. A. 2010a. Do males offset the cost of female aggression? An experimental test in a biparental songbird. *Behavioral Ecology*, 21, 161–8.

ROSVALL, K. A. 2010b. A novel cost of a sexually selected trait in females: more aggressive female tree swallows incubate less. *Integrative and Comparative Biology*, 50, E152.

ROSVALL, K. A. 2011a. By any name, female-female competition yields differential mating success. *Behavioral Ecology*, 22, 1144–6.

ROSVALL, K. A. 2011b. Intrasexual competition in females: evidence for sexual selection? *Behavioral Ecology*, 22, 1131–40.

ROSVALL, K. A., BURNS, C. M. B., BARSKE, J., GOODSON, J. L., SCHLINGER, B. A., SENGELAUB, D. R., & KETTERSON, E. D. 2012. Neural sensitivity to sex steroids predicts individual differences in aggression: implications for behavioural evolution. *Proceedings of the Royal Society of London—Series B: Biological Sciences*, 279, 3547–55.

ROSVALL, K. A. & PETERSON, M. P. 2014. Behavioral effects of social challenges and genomic mechanisms of social priming: what's testosterone got to do with it? *Current Zoology*, 60, 791–803.

RUBENSTEIN, D. R. 2012. Sexual and social competition: broadening perspectives by defining female roles. *Philosophical Transactions of the Royal Society B—Biological Sciences*, 367, 2248–52.

SCAIA, M. F., MORANDINI, L., NOGUERA, C. A., RAMALLO, M. R., SOMOZA, G. M., & PANDOLFI, M. 2018. Fighting cichlids: dynamic of intrasexual aggression in dyadic agonistic encounters. *Behavioural Processes*, 147, 61–9.

SCARPONI, V., CHOWDHURY, D., & GODIN, J. G. J. 2015. Male mating history influences female mate choice in the Trinidadian guppy (*Poecilia reticulata*). *Ethology*, 121, 1091–103.

SCARPONI, V. & GODIN, J. G. J. 2018. Female assessment of male functional fertility during mate choice in a promiscuous fish. *Ethology*, 124, 196–208.

SCHEUNEMANN, L., LAMPIN-SAINT-AMAUX, A., SCHOR, J., & PREAT, T. 2019. A sperm peptide enhances long-term memory in female *Drosophila*. *Science Advances*, 5, eaax3432.

SCHMITT, D. P. & SHACKELFORD, T. K. 2003. Nifty ways to leave your lover: the tactics people use to entice and disguise the process of human mate poaching. *Personality and Social Psychology Bulletin*, 29, 1018–35.

SEYFARTH, R. M. 1976. Social relationships among adult female baboons. *Animal Behaviour*, 24, 917–38.

SHARP, S. P. & CLUTTON-BROCK, T. H. 2011a. Competition, breeding success and ageing rates in female meerkats. *Journal of Evolutionary Biology*, 24, 1756–62.

SHARP, S. P. & CLUTTON-BROCK, T. H. 2011b. Reluctant challengers: why do subordinate female meerkats rarely displace their dominant mothers? *Behavioral Ecology*, 22, 1337–43.

SHIMOJI, H., AONUMA, H., MIURA, T., TSUJI, K., SASAKI, K., & OKADA, Y. 2017. Queen contact and among-worker interactions dually suppress worker brain dopamine as a potential regulator of reproduction in an ant. *Behavioral Ecology and Sociobiology*, 71, 35.

SICOTTE, P., TEICHROEB, J. A., VAYRO, J. V., FOX, S. A., BĂDESCU, I., & WIKBERG, E. C. 2017. The influence of male takeovers on female dispersal in *Colobus vellerosus*. *American Journal of Primatology*, 79, e22436.

SILK, J. B., ALBERTS, S. C., & ALTMANN, J. 2004. Patterns of coalition formation by adult female baboons in Amboseli, Kenya. *Animal Behaviour,* 67, 573–82.

SQUIRES, Z. E., WONG, B. B. M., NORMAN, M. D., & STUART-FOX, D. 2015. Last male sperm precedence in a polygamous squid. *Biological Journal of the Linnean Society,* 116, 277–87.

STOCKLEY, P. & BRO-JORGENSEN, J. 2011. Female competition and its evolutionary consequences in mammals. *Biological Reviews,* 86, 341–66.

STRAUSS, E. D. & HOLEKAMP, K. E. 2019. Social alliances improve rank and fitness in convention-based societies. *Proceedings of the National Academy of Sciences of the United States of America,* 116, 8919–24.

STUART, R., FRANCOEUR, A., & LOISELLE, R. 1987. Lethal fighting among dimorphic males of the ant, *Cardiocondyla wroughtonii. Naturwissenschaften,* 74, 548–9.

SURBECK, M., BOESCH, C., GIRARD-BUTTOZ, C., CROCKFORD, C., HOHMANN, G., & WITTIG, R. M. 2017. Comparison of male conflict behavior in chimpanzees (*Pan troglodytes*) and bonobos (*Pan paniscus*), with specific regard to coalition and post-conflict behavior. *American Journal of Primatology,* 79, e22641.

SWIERK, L., TENNESSEN, J. B., & LANGKILDE, T. 2015. Sperm depletion may not limit male reproduction in a capital breeder. *Biological Journal of the Linnean Society,* 116, 684–90.

TINBERGEN, N. 1936. The function of sexual fighting in birds; and the problem of the origin of "territory." *Bird-banding,* 7, 1–8.

TOBIAS, J. A., MONTGOMERIE, R., & LYON, B. E. 2012. The evolution of female ornaments and weaponry: social selection, sexual selection and ecological competition. *Philosophical Transactions of the Royal Society B—Biological Sciences,* 367, 2274–93.

VAN SCHAIK, C. P., PANDIT, S. A., & VOGEL, E. R. 2006. Toward a general model for male-male coalitions in primate groups. *In:* VAN SCHAIK, C. P. & KAPPELER, P. M. (eds.) *Cooperation in Primates and Humans.* Amsterdam: Springer.

VENNER, S., BERNSTEIN, C., DRAY, S., & BEL-VENNER, M. C. 2010. Make love not war: when should less competitive males choose low-quality but defendable females? *American Naturalist,* 175, 650–61.

WANKER, R. & FISCHER, J. 2001. Intra- and interindividual variation in the contact calls of spectacled parrotlets (*Forpus conspicillatus*). *Behaviour,* 138, 709–26.

WANKER, R., SUGAMA, Y., & PRINAGE, S. 2005. Vocal labelling of family members in spectacled parrotlets, *Forpus conspicillatus. Animal Behaviour,* 70, 111–8.

WASSER, S. K. & BARASH, D. P. 1983. Reproductive suppression among female mammals: implications for biomedicine and sexual selection theory. *The Quarterly Review of Biology,* 58, 513–38.

WATSON, N. L. & SIMMONS, L. W. 2010. Reproductive competition promotes the evolution of female weaponry. *Proceedings of the Royal Society of London—Series B: Biological Sciences,* 277, 2035–40.

WHITE, D. J. & GALEF JR., B. G. 1999. Social effects on mate choices of male Japanese quail, *Coturnix japonica. Animal Behaviour,* 57, 1005–12.

WONG, B. B. M. & ROSENTHAL, G. G. 2006. Female disdain for swords in a swordtail fish. *American Naturalist,* 167, 136–40.

CHAPTER 9

Synthesis and Outlook

9.1 Brief outline of the chapter

In this final chapter I want to briefly recap what I presented in the previous chapters and provide a few ideas of what might be done in the future to move the field forward. All three factors I discussed as relevant in male mate choice, male investment in reproduction, sex ratios, and variability in partner quality, are still emerging fields in sexual selection research and need more theoretical and empirical work. I suggest that variability in female quality is more important and more complex than currently known. The same is true for sex ratios. On the other hand, I suggest that sheer investment in gametes may be a little less important than currently assumed. Most importantly we need to explore the interactions of these three pathways to male mate choice. Female competition and also female ornamentation are still somewhat enigmatic and both topics are likely to grow in importance for our understanding of sexual selection. I think considering male and female choice together, as well as female and male competition will ultimately provide a more complete picture of Darwinian sexual selection.

9.2 Introduction

In this book I tried to cover and summarize some of what we know about two additional aspects of sexual selection that in my view are critical to our understanding of it. I suggest that we do not know enough about male mate choice, and that the knowledge we have is still patchy and in need of more good theory. There are some excellent theoretical studies, but more is needed to guide and evaluate empirical research (Courtiol et al., 2016; Fitzpatrick and Servedio, 2017, 2018). We have a

very detailed understanding of female choice, and a simple first step would be to apply some of the same paradigms we use in female mate choice for advancing research into male mate choice.

Historically, female mate choice was ignored at the expense of male competition (Milam, 2010). This situation has now flipped and a majority of studies in sexual selection focus on female choice (Chapter 1). I think more studies on male competition would also provide a better framework for understanding competition in females. It seems clear that female competition comes in different forms as compared with male competition, but it is not necessarily less important. I contend that female aggression has been neglected in sexual selection and mainly studied in contexts other than sexual selection, such as maternal care.

Closely related to the two topics of female competition and male mate choice is the presence of ornamental traits in females (Chapter 7). We are starting to understand what role they might play in evolution, but the picture is far from clear. Furthermore, what we know is often taxonomically biased: female ornaments are relatively well described in birds (Doutrelant et al., 2020), with some interesting experimentation in lizards, but beyond this, there is little knowledge. The same is also true for male mate choice. Even considering the many examples compiled in Chapter 3 we know very little, especially compared with the many studies on female mate choice. Indeed, the field is so small that individual groups or people leave important marks: Paul Verrell, to name just one, almost single-handedly conducted the studies that led to our knowledge about male mate choice in salamanders (Verrell, 1985, 1986, 1989, 1994, 1995). Furthermore, female song is just now being

Male Choice, Female Competition, and Female Ornaments in Sexual Selection. Ingo Schlupp, Oxford University Press (2021). © Ingo Schlupp (2021).
DOI: 10.1093/oso/9780198818946.003.0009

recognized as important in birds (Riebel, 2003), with limited knowledge of many other taxa where males vocalize. I expect this phenomenon to be important in other groups as well, including anurans, some insects, and even fishes.

How can we move forward? One of the most powerful frameworks to organize hypotheses in biology was created by Niko Tinbergen in 1963 (Tinbergen, 1963). He outlines four questions that should to be asked by biologists, all of which are still relevant and widely applied. In relation to male mate choice we have partial answers to one of them. But all four questions need to become the focus of more theoretical and empirical research. The second of Tinbergen's questions, the one about "survival value," is the one this book is mainly concerned with. We now have some theory and empirical data on how male mate choice, female competition, and female ornaments can be adaptive. We know little, however, about what the mechanisms for male mate choice are. For example, in female choice we have a fairly good picture of the hormonal and neurobiological mechanisms (DeAngelis and Hofmann, 2020); this is not the case for male mate choice. Can we assume the same mechanisms are involved? How sex specific are hormones, really (Lipshutz et al., 2019)? Equally, we know little about the evolutionary history of male mate choice or its ontogenetic development. Is it really almost universal, as I suggest? Is it a byproduct of female choice? Although I am a bit critical of this idea, it is a viable hypothesis that needs more tests, both for a genetic correlation of male and female choosiness, and as an explanation for female ornaments. The logic might also apply to aggressive behavior in males and females. The framework proposed by Tinbergen (1963) and his four questions can guide future research.

In addition to more theory we also need to reevaluate some of our methods. Methods for studying animal behavior have not changed much over the past few decades and rely heavily on direct observation of animal interactions. This is changing with the data revolution and the advance of AI and video-based machine learning. Just one example is the software DeepLabCut (Mathis et al., 2018; Nath et al., 2019), which uses machine learning to train a neural network to identify subtle behavioral differences. More advances or changes like this can be expected.

From comparisons of binary choice tests and absolute preference functions it is clear that how we experimentally measure preferences can lead to different findings. I have already mentioned that the operational sex ratio (OSR) used in most binary choice test is typically 2:1. This probably gives us a distorted view of the importance of OSR in mating decisions. More importantly, preferences and mating decisions can be highly individualized, yet we typically consider them as population means. The established two main factors driving female choice are direct benefits and indirect, genetic benefits. This predicts great concordance of preferences, but there is an alternative. In some cases, we know that mate choice is for the best fit in a mating partner, and for those cases population means may not be the best descriptor of mate choice. This is highlighted in studies of mate choice for a matching partner, e.g. based on major histocompatibility complex (MHC), where the best match depends on the chooser's own MHC, not a mean preference for a certain trait.

Also, such individualized preferences can of course interact either synergistically or antagonistically with population-wide preferences. For a male, a small female may be a better match in terms of MHC, but he may sire many more offspring with a larger female. What is the best decision in this scenario? Such tradeoffs deserve more experimental work and theoretical consideration.

It is now abundantly clear that the social environment and social evolution (Araya-Ajoy et al., 2020) is of key importance for mate choice (Scauzillo and Ferkin, 2019). I have already highlighted the importance of the social environment several times. Many studies are now taking this into account, but almost all are doing so from a female choice perspective. There are very few examples of looking at male mate choice in a social context (Schlupp and Ryan, 1997). Also, female competition is rarely studied in a social context (Makowicz et al., 2020). More studies along those lines would be useful.

Fortunately, there are quite a few animal mating systems where we can easily test new hypotheses. In sticklebacks, for example, male and female mate choice is simultaneous and male choosiness is based on female fecundity (Kraak and Bakker, 1998). In this species we also have a good idea of the role of MHC in female mate choice. In several lizard species we have good data on male mate choice and its relationship with female ornaments. And in some amphibians and fishes, we know that males choose although they make no investment in the offspring (Schlupp, 2018). Such species would be good candidates to study the role of female quality in male choice. A large number of studies currently conclude that fecundity is behind male choice. Arguably, fecundity is extremely important and may even be the most important direct benefit to males, but it also is clearly not the only one. Why, for example, would males mate with a parasitized or sick female? We know that females avoid parasitized males, but more studies on male choice would be useful.

9.3 Male mate choice is common

It turns out that male mate choice is relatively common—once we start looking for it. It is even found in species that have no paternal investment and a male-biased sex ratio (Schlupp, 2018). It seems that in these cases differences in female quality alone can lead to the expression of male mate choice. Nonetheless, male mate choice still seems like a somewhat overlooked phenomenon. We have some good theory for the evolution of male mate choice and the related female competition and female ornaments, and limited but solid empirical evidence that needs to be interpreted in the light of new theory. For me a few important things emerge and I have tried to organize the chapters of this book roughly along those lines. The first thing to look at more closely is the role of investment in gametes and sex roles and the general characteristics we ascribe them. I suggest that we should consider them to be less rigid than currently often portrayed. Clearly the traditional approach introduced by Darwin that males compete and females choose is correct, but the idea that these are the *only* two components in sexual selection might be wrong. Culturally, the traditional view of sex roles has somewhat dominated the way we ask questions in science. Yet there are additional, new questions to be asked to match the complexity we actually find in nature. Certainly, too,

this complexity should be reflected in the textbooks we use to train our students. Furthermore, and troubling to me, the traditional view of sex roles has permeated into human culture and the image of coy women and pushy men is as ubiquitous as it is simplistic. I worry that such an incomplete view is sometimes used to excuse unacceptable bad human behavior in a "sciency" sounding way. Sex roles might, for example, provide excuses for the maintenance of patriarchal systems, where "science" may then be used to justify the oppression of women. If we modify our understanding of sex roles, we can rethink stereotypes like "boys will be boys" or "women are coy." The traditional view seems to take away agency and responsibility from both men and women. I think that mating interactions should not be reduced to the sex roles proposed by Bateman (1948) and Trivers (1972). After all, Darwin developed his ideas in the Victorian era, and it seems likely that the prevailing view of gender roles at that time influenced not only his own thinking, but also how his ideas were perceived and when and how they were translated into scientific questions. I also hope that we will explore how binary sex really is in the animal kingdom. We are realizing that in humans gender is not completely binary and that gives us tools and a framework for exploring this in all of the animal kingdom.

History aside, there are a few interesting patterns that arise from the studies reviewed in this book. When males choose, they seem to be choosing mainly based on direct benefits, especially fecundity—or so we think based on the studies we have. But females differ in many more characteristics than just fecundity. Based on the long list of traits that females use in mate choice, it would be interesting to look at similar traits and their role in male mate choice (Table 9.1). I think evidence for male choice for indirect benefits is absent, probably difficult to collect, but studies—both theoretical and empirical—on this subject would be extremely interesting.

Finally, we know that female choice has a number of effects outside of sexual selection *sensu stricto*. For example, sexual selection seems to have an important role in speciation. Could this also be said for male mate choice? Can it drive speciation or other important processes?

Table 9.1 Predicted male preferences for female traits.

Female trait	Predicted male preference
Fecundity	Larger females preferred
Female resources	Females with more resources preferred
Health	Healthy and parasite-free females preferred
Condition	Well-fed females preferred
Age	Generally younger females preferred; senescing females rejected
Parenting	Females that invest in offspring (good mothers) preferred
Experience	More experienced females preferred
Compatibility	Genetically (major histocompatibility complex) or socially compatible females preferred

I devoted three chapters to the likely underlying causes of male mate choice. These are differences in female quality, male investment in reproduction, and OSR. All three causes are predicted to have some influence on male mate choice. Most importantly I think they can interact in interesting ways and have synergistic or antagonistic effects on mate choice. For example, in bush crickets, strong male investment in large spermatophores forces some males to drop out of reproduction in adverse environmental conditions (Gwynne and Simmons, 1990). This in turn can lead to a female-biased OSR. Together this favors the expression of male mate choice.

9.4 Differences in partner quality

We have a very detailed understanding of female choice. Both the mechanisms and evolutionary consequences of female choice are well documented. Compared with that, male mate choice is yet poorly understood. We know, however, from female choice that many traits can be the target of female preferences. Similar studies on male mate choice are often lacking. A majority of male choice studies are detecting fecundity as the direct benefit for males. This may be owing to some bias in the published literature, but may also be the best pathway for male mate choice to evolve (Servedio and Lande, 2006).

Males may also choose for other reasons. In particular, male mate choice for compatible females,

either genetically or socially, is not well known. Is this because it is truly rare, or because we have not been looking for it yet? I suspect it is the latter. Table 9.1 provides a few ideas of what might be fruitful to investigate. For example, predicting male preferences for female ornamentation is not completely straightforward. If the ornament predicts female quality in any way, males should prefer more ornamented females, but if the ornament is present owing to a genetic correlation, and provides no valuable information to the choosing male, the more inconspicuous female should be preferred because consorting with her means a lower risk of predation. Luckily, these questions can be addressed empirically. Finally, males should also prefer females that are compatible mates. This can be either genetic compatibility (via MHC) or social compatibility, leading to stronger duos that can acquire more or better resources. If this is the case, both sexes should choose, and their mutual choices would be in the same direction. The results of this could look very much like assortative mating. This should be important for species with some form of monogamy. There is a rich literature on costs and benefits of monogamy and also on the mechanisms regulating it. There are two especially important aspects to that. First, often pairs get better over time at raising offspring. This suggests that they learn about their environments and avoid mistakes, but they may also become more compatible with each other. Showing compatibility might be the function of duetting in birds, for example. What this means, though, is that mate choice for compatible partners is very different from mate choice for the largest or most aesthetically pleasing mate and may lead to completely different mating patterns.

Equally important is a closer look at the cost of male mate choice (Härdling and Kokko, 2005). The cost may be holding back the evolution of female competition and ornamentation, but how costly is male mate choice really? It seems the biggest cost to males is missed opportunities as they reject certain females. With mating opportunities generally being rare for males, this may be a substantial cost. Also, can males reap indirect benefits like females? Does something like a "sexy daughter" effect exist? Do choosy males not only have more daughters but also better-quality ones?

However, this may not be true for high-quality males that have multiple mating opportunities. They may face a very low cost of choice and may be selected to carefully allocate their potentially limited mating resources, ranging from sperm to nuptial gifts and beyond. This idea has been modeled using game theory (Härdling and Kokko, 2005; Härdling et al., 2008) showing that higher male cost for sperm can lead to males and females being more choosy.

Finally, one interesting consequence of male choosiness may be that, contrary to the standard paradigm, not all females find a mate. In insects the rate of lifelong virginity of females is around 5% (Rhainds, 2019), but clearly this deserves more attention. Rates of lifelong virginity for males are, of course, thought to be much higher.

9.5 Investment in reproduction

Based on investment in the gametes alone, females are predicted to be choosy. But males, which invest far less in their gametes, can invest in their partners and offspring in other forms. Even if they do not catch up completely, they may narrow the gap and, as they do so, they should benefit from being selective about with whom they breed. Furthermore, as we already discussed, even in cases where males invest very little they may be selected to be choosy because of differences in female quality or availability. What seems to be puzzling at first can be resolved once we step away from the traditional view of sex roles. Again, another consideration is the role of sex roles. This concept is based on investment of males and females in their gametes and has been used to broadly predict behavioral patterns of choosy females and promiscuous males. Those patterns have been studied often and have mostly been confirmed. But has this prevented us from asking questions beyond the traditional sex roles? Do we know so little about female competition and male mate choice because the Bateman–Trivers concept of sex roles has dominated the field and influenced the questions to be asked? Are Bateman gradients maybe only predictive of the strength of choice, not binary roles? In an opinion paper, Cain and Rosvall argue that a conceptual weakness of Bateman gradients is that they ignore ecology and access to critical resources (Cain and Rosvall, 2014). The authors advocate for a better understanding of the role of ecological resources and social evolution in female competition.

If both sexes have preferences and exercise some kind of choosiness, mate choice becomes mutual (Johnstone et al., 1996; Johnstone, 1997; Bergstrom and Real, 2000). This can for example be on an individual level, as if partners negotiate an agreement (Bergstrom and Real, 2000), but can also support speciation (Lande et al., 2001). Although there are many examples of mutual mate choice the relationship with only male or female choice is not clear (Figure 9.1), and mutual mate choice is actually not that common (Kokko and Johnstone, 2002)—or not that often recognized. In gobies, for example, it seems that the OSR is a driving force of the mating system, but it is mainly female choice early in the season, when males are abundant and then male choice later in the season, when males are rare (Amundsen, 2018). Both sexes are choosy, but at different times. Is this still a system with mutual mate choice?

Maybe a good example is mutual mate choice in humans. In our own species we can be fairly certain that women and men both have preferences, which tend to be complex and may or may not be important in actual matings (Courtiol et al., 2010). Also, they may pay attention to somewhat different traits, but each gender is trying to mate with the best possible mate. Mate preferences for height are a good example. Here, differences in preferences tend to lead to a compromise (Sendova-Franks, 2013). What are the relative roles of conflict and cooperation in

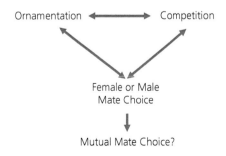

Figure 9.1 Potential interaction of female choice and male choice, when it occurs simultaneously. The resulting mutual choice is the result of the interaction of both.

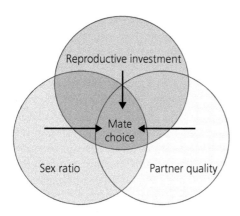

Figure 9.2 Interaction of sex ratio, reproductive investment, and partner quality. The center field is where mate choice is strongest.

finding mutually agreeable mating partners? Of course, humans are a somewhat difficult example as culture plays such an important role in our mating system.

Clearly, mutual mate choice is key to understanding the simultaneous nature of male and female mate choice. How do we view and model the interactions that lead to matings and reproduction (Figure 9.2)? What are the processes that lead to a situation where both partners (females and males, under the old paradigm) have agency and input into the outcome of the mating interaction? How much is this governed by sexual conflict and social evolution? What are the relative roles of male and female choice and how do they lead to mutual mate choice (Figure 9.3).

9.6 Sex ratios

The conditions that lead to male mate choice are heavily driven by ecology including the ratio of choosers to courters. This ratio—either captured as OSR or adult sex ratio (ASR)—is a dynamic characteristic of populations, which predicts that male mate choice should be shown under some conditions, but not others. Choosiness varies with OSR, as demonstrated by several examples in Chapter 6, including in gobies. Reflecting the OSR, female choosiness is high early in the breeding season and then eventually flips toward the end of the breeding season when males are rare. Such conditions are likely not limited to gobies, and more examples

would be good to clarify the generality of this phenomenon. I imagine there will be more examples once we start looking more carefully. Variation in sex ratios and the associated male mate choice can play out at various levels either on the population or the species level. Based on breeding ecology, male mate choice can be found on the population level as shown, for example, in the peacock blenny, *Salaria pavo* (Almada et al., 1994, 1995) (Chapter 6). Male mate choice and female competition are found in the populations of the Portuguese Algarve because breeding sites are rare. What would happen if we experimentally provided ample breeding opportunities for this population? Likely they would flip to female choice and male competition. Besides, the Algarve is a wonderful area and field work there is great fun! The opposite experiment could be done in a habitat where breeding sites are abundant. These populations might be in the "Area of behavioral flexibility" depicted in Figure 5.1. Sex ratios can also be relevant to the evolution of male mate choice on the species level as suggested for sex-role reversed shore birds (Liker et al., 2013). Finally, as seen in the example of Scandinavian gobies (Forsgren et al., 2004; Amundsen, 2018), (Chapter 6), males and females may switch from being choosy to competing based on the seasonally changing OSR. Together this tells us again that neither male nor female choice are as fixed a strategy as we currently think. I argue that choosiness based on sex ratios may vary on the level of the individual, where some males that are perceived as "high-quality" or are especially "attractive" may be selective, while other males may not. A similar argument for individual-based mating decisions and flexible switches between behaviors was already made before, for example, by Ah-King and Gowaty (2016). It seems important to understand the underlying individual costs and benefits of choosiness to predict when to expect expression of this trait. More generally, I think the role of the individual is underestimated in behavior, where the focus still lies on characterizing population means.

Finally, I am wondering about the potential benefits of preferences for socially compatible mates. As in preferences for a matching MHC, there may be a good partner available for any given individual based on compatibility, not uniform preferences for

Figure 9.3 Visual model of how male and female choice might interact.

a "best" mate. This phenomenon may come as assortative mating for certain traits, like size, or—in humans—for socioeconomic stratum, but would also predict that certain behavioral types (or personalities) are a better match for some individuals than for others. Behavioral types have been discussed for a long time and have been implicated as important traits in mate choice; often not in relation to compatibility, but rather as a complex indicator trait in female choice (Munson et al., 2020).

I have not discussed mechanisms very much. This is not because I think that mechanisms are unimportant. I strongly believe that the genetic and epigenetic basis of behavior, neurobiology, and hormones are actually incredibly important in understanding mate choice both in females and males. My hope is that as we make progress in theory and empirical studies of the adaptive value of male mate choice, colleagues will start investigating more of the mechanistic side, so that we can develop research programs true to Tinbergen's four questions (Tinbergen, 1963). One could even ask if with the view of, for example, sex hormones as sex specific (Lipshutz et al., 2019), we unnecessarily maintain a gendered view of mechanisms underlying behavior. In reality we now know that the role of testosterone is highly complex and very important in both males and females.

9.7 Female competition

Understanding female competition seems even more difficult. Females clearly compete with each other, but it is less clear if they compete directly for males in the way males compete for females. Instead it seems that in general females compete more indirectly for resources associated with males. In a way the distinction between those two ways of female competition

seems moot, as it leads to a similar outcome. But if indirect competition is prevalent in females, we would not predict the many traits associated with direct, active, and combat-like competition found in males. Similarly, dual-function traits that serve as ornaments and armaments are predicted to be less common. Finally, indirect competition may also help explain the relative paucity of evolved female ornaments that might be signals to other females. Alternatively, females may advertise their quality to males with traits that are less conspicuous than the corresponding traits in males. Here I suspect that much more research is needed to understand the mechanisms and evolutionary pathways of female ornamentation. The field of social evolution is probably going to make substantial contributions here (Lyon and Montgomerie, 2012).

9.8 Female ornamentation

Ornamental traits in females have been hiding in plain sight for some time. Visual ornaments are best studied in birds (Doutrelant et al., 2020) and more work is needed looking for such ornaments in other taxa. Female song as an ornamental trait is also an emerging field, again with most data available for birds (Riebel, 2003; Odom et al., 2014). Lizards are a group in which ornaments are also relatively common and in lizards some experimental work has been done, for example on fence lizards, *Sceloporus* sp. (Assis et al., 2018). There are many open questions associated with female ornaments. Are they really all ornaments? Are they a byproduct of selection on male ornaments? The latter seems unlikely to me, but this is nonetheless a reasonable hypothesis. If they are ornaments, what role do they play in male choice and female competition? Why are there relatively few armaments in females? It is also interesting that female ornaments (and female competition) do not resemble male ornaments and competition. While males may risk injuries and their life during aggressive encounters, that is not common in females. Intuitively, that makes sense based on their investment in large gametes. Also, whether much mortality through predators is associated with female ornaments is not clear.

9.9 Short summary

Male mate choice and female competition are the missing parts to a better understanding of sexual selection. They are the counterparts to the well-understood female choice and male competition. They are similar in some regards, but also different in many ways. They present challenges for more theoretical and empirical work, but more importantly exciting opportunities.

9.10 References

AH-KING, M. & GOWATY, P. A. 2016. A conceptual review of mate choice: stochastic demography, within-sex phenotypic plasticity, and individual flexibility. *Ecology and Evolution,* 6, 4607–42.

ALMADA, V. C., GONÇALVES, E. J., OLIVEIRA, R. F., & SANTOS, A. J. 1995. Courting females: ecological constraints affect sex roles in a natural population of the blenniid fish *Salaria pavo. Animal Behaviour,* 1995, 1125–7.

ALMADA, V. C., GONÇLALVES, E. J., SANTOS, A. J., & BAPTISTA, C. 1994. Breeding ecology and nest aggregations in a population of *Salaria pavo* (Pisces: Blenniidae) in an area where nest sites are very scarce. *Journal of Fish Biology,* 45, 819–30.

AMUNDSEN, T. 2018. Sex roles and sexual selection: lessons from a dynamic model system. *Current Zoology,* 64, 363–92.

ARAYA-AJOY, Y. G., WESTNEAT, D. F., & WRIGHT, J. 2020. Pathways to social evolution and their evolutionary feedbacks. *Evolution,* 74, 1894–907.

ASSIS, B. A., SWIERK, L., & LANGKILDE, T. 2018. Performance, behavior and offspring morphology may offset reproductive costs of male-typical ornamentation for female lizards. *Journal of Zoology,* 306, 235–42.

BATEMAN, A. J. 1948. Intra-sexual selection in *Drosophila. Heredity,* 2, 349–368.

BERGSTROM, C. T. & REAL, L. A. 2000. Towards a theory of mutual mate choice: lessons from two-sided matching. *Evolutionary Ecology Research,* 2, 493–508.

CAIN, K. E. & ROSVALL, K. A. 2014. Next steps for understanding the selective relevance of female-female competition. *Frontiers in Ecology and Evolution,* 2.

COURTIOL, A., ETIENNE, L., FERON, R., GODELLE, B., & ROUSSET, F. 2016. The evolution of mutual mate choice under direct benefits. *American Naturalist,* 188, 521–38.

COURTIOL, A., RAYMOND, M., GODELLE, B., & FERDY, J. B. 2010. Mate choice and human stature: homogamy as a unified framework for understanding mating preferences. *Evolution,* 64, 2189–203.

DEANGELIS, R. S. & HOFMANN, H. A. 2020. Neural and molecular mechanisms underlying female mate choice decisions in vertebrates. *Journal of Experimental Biology,* 223, jeb207324.

DOUTRELANT, C., FARGEVIEILLE, A., & GRÉGOIRE, A. 2020. Evolution of female coloration: what have we learned from birds in general and blue tits in particular. *In:* NAGUIB, M., BARRETT, L., HEALY, S. D., PODOS, J., SIMMONS, L. W., & ZUK, M. (eds.) *Advances in the Study of Behavior.* London: Academic Press.

FITZPATRICK, C. L. & SERVEDIO, M. R. 2017. Male mate choice, male quality, and the potential for sexual selection on female traits under polygyny. *Evolution,* 71, 174–83.

FITZPATRICK, C. L. & SERVEDIO, M. R. 2018. The evolution of male mate choice and female ornamentation; a review of mathematical models *Current Zoology,* 64, 323–33.

FORSGREN, E., AMUNDSEN, T., BORG, Å. A., & BJELVENMARK, J. 2004. Unusually dynamic sex roles in a fish. *Nature,* 429, 551.

GWYNNE, D. & SIMMONS, L. 1990. Experimental reversal of courtship roles in an insect. *Nature,* 346, 172–4.

HÄRDLING, R., GOSDEN, T., & AGUILEE, R. 2008. Male mating constraints affect mutual mate choice: prudent male courting and sperm-limited females. *American Naturalist,* 172, 259–71.

HÄRDLING, R. & KOKKO, H. 2005. The evolution of prudent choice. *Evolutionary Ecology Research,* 7, 697–715.

JOHNSTONE, R. A. 1997. The tactics of mutual mate choice and competitive search. *Behavioral Ecology & Sociobiology,* 40, 51–9.

JOHNSTONE, R. A., REYNOLDS, J. D., & DEUTSCH, J. C. 1996. Mutual mate choice and sex differences in choosiness. *Evolution,* 50, 1382–91.

KOKKO, H. & JOHNSTONE, R. A. 2002. Why is mutual mate choice not the norm? Operational sex ratios, sex roles and the evolution of sexually dimorphic and monomorphic signalling. *Philosophical Transactions of the Royal Society of London B—Biological Sciences,* 357, 319–30.

KRAAK, S. B. M. & BAKKER, T. C. M. 1998. Mutual mate choice in sticklebacks: attractive males choose big females, which lay big eggs. *Animal Behaviour,* 56, 859–66.

LANDE, R., SEEHAUSEN, O., & VAN ALPHEN, J. J. 2001. Mechanisms of rapid sympatric speciation by sex reversal and sexual selection in cichlid fish. *Genetica,* 112, 435–43.

LIKER, A., FRECKLETON, R. P., & SZEKELY, T. 2013. The evolution of sex roles in birds is related to adult sex ratio. *Nature Communications,* 4, 1–6.

LIPSHUTZ, S., GEORGE, E., BENTZ, A., & ROSVALL, K. 2019. Evaluating testosterone as a phenotypic integrator:

from tissues to individuals to species. *Molecular and cellular endocrinology*, 496, 110531.

LYON, B. E. & MONTGOMERIE, R. 2012. Sexual selection is a form of social selection. *Philosophical Transactions of the Royal Society B—Biological Sciences*, 367, 2266–73.

MAKOWICZ, A., MURRAY, L., & SCHLUPP, I. 2020. Size, species and audience type influence heterospecific female–female competition. *Animal Behaviour*, 159, 47–58.

MATHIS, A., MAMIDANNA, P., CURY, K. M., ABE, T., MURTHY, V. N., MATHIS, M. W., & BETHGE, M. 2018. DeepLabCut: markerless pose estimation of user-defined body parts with deep learning. *Nature Neuroscience*, 21, 1281–9.

MILAM, E. L. 2010. *Looking for a Few Good Males*. Baltimore, MD: The Johns Hopkins University Press.

MUNSON, A. A., JONES, C., SCHRAFT, H., & SIH, A. 2020. You're just my type: mate choice and behavioral types. *Trends in Ecology & Evolution*, 35, 823–33.

NATH, T., MATHIS, A., CHEN, A. C., PATEL, A., BETHGE, M., & MATHIS, M. W. 2019. Using DeepLabCut for 3D markerless pose estimation across species and behaviors. *Nature Protocols*, 14, 2152–76.

ODOM, K. J., HALL, M. L., RIEBEL, K., OMLAND, K. E., & LANGMORE, N. E. 2014. Female song is widespread and ancestral in songbirds. *Nature Communications*, 5, 3379.

RHAINDS, M. 2019. Ecology of female mating failure/lifelong virginity: a review of causal mechanisms in insects and arachnids. *Entomologia Experimentalis et Applicata*, 167, 73–84.

RIEBEL, K. 2003. The"mute" sex revisited: vocal production and perception learning in female songbirds. *Advances in the Study of Behavior*, 33, 49–86.

SCAUZILLO, R. C. & FERKIN, M. H. 2019. Factors that affect non-independent mate choice. *Biological Journal of the Linnean Society*, 128, 499–514.

SCHLUPP, I. 2018. Male mate choice in livebearing fishes: an overview. *Current Zoology*, 64, 393–403.

SCHLUPP, I. & RYAN, M. J. 1997. Male sailfin mollies (*Poecilia latipinna*) copy the mate choice of other males. *Behavioral Ecology*, 8, 104–7.

SENDOVA-FRANKS, A. 2013. human mutual mate choice for height results in a compromise. *Animal Behaviour*, 86, 2–2.

SERVEDIO, M. R. & LANDE, R. 2006. Population genetic models of male and mutual mate choice. *Evolution*, 60, 674–85.

TINBERGEN, N. 1963. On aims and methods of ethology. *Zeitschrift für Tierpsychologie.*, 20, 410–433.

TRIVERS, R. (ed.) 1972. *Parental Investment and Sexual Selection*. Chicago, IL: Aldine.

VERRELL, P. A. 1985. Male mate choice for large, fecund females in the red-spotted newt, *Notophthalmus viridescens*: how is size assessed? *Herpetologica*, 1985, 382–6.

VERRELL, P. A. 1986. Male discrimination of larger, more fecund females in the smooth newt, *Triturus vulgaris*. *Journal of Herpetology*, 1986, 416–22.

VERRELL, P. A. 1989. Male mate choice for fecund females in a plethodontid salamander. *Animal Behaviour*, 38, 1086–8.

VERRELL, P. A. 1994. Males may choose larger females as mates in the salamander *Desmognathus fuscus*. *Animal Behaviour*, 47, 1465–7.

VERRELL, P. A. 1995. Males choose larger females as mates in the salamander *Desmognathus santeetlah*. *Ethology*, 99, 162–71.

Index